International Conference on Power and Propulsion Alternatives for Ships 2017

Rotterdam, The Netherlands
8 November 2017

ISBN: 978-1-5108-8304-8

Printed from e-media with permission by:

Curran Associates, Inc.
57 Morehouse Lane
Red Hook, NY 12571

The Royal Institution of Naval Architects
8-9 Northumberland Street
London, WC2N 5DA
United Kingdom

Phone: 020 7235 4622
Fax: 020 7259 5912

publications@rina.org.uk

Additional copies of this publication are available from:

Curran Associates, Inc.
57 Morehouse Lane
Red Hook, NY 12571 USA
Phone: 845-758-0400
Fax: 845-758-2633
Email: curran@proceedings.com
Web: www.proceedings.com

CONTENTS

Power and Propulsion Alternatives for Ships

8 November 2017, Rotterdam, The Netherlands

PAPERS

THE ROYAL INSTITUTION OF NAVAL ARCHITECTS
8-9 Northumberland Street, London, WC2N 5DA, Telephone: +44 (0)20 7235 4622

CURRENT AND FUTURE SCALE LIMITATION FOR ALTERNATIVE MARINE POWER AND PROPULSION SOLUTIONS

J-E Räsänen, Foreship Ltd., Finland

SUMMARY

Alternative energy sources for marine power and propulsion are in a phase of development which may be decisive. For several years, the potential benefits to shipping offered by battery energy storage have faced accusations of hype, focusing on whether the footprint of plant required can truly justify the energy generated. Alternative power sources play a significant role in the topic of Autonomous ships and how these would be powered, to minimize maintenance requirements onboard.

It seems obvious that fuel cells will, or at least potentially can, play a significant role as an alternative marine power source. From Foreship's stand point, the future of fuel cells is highly dependent on scaling up the right technology and improving efficiency to a level significantly higher than that of a combustion engine.

ABBREVIATIONS

Li-NMC	Lithium nickel manganese cobalt oxide battery ($LiNiMnCoO_2$)
Li-LFP	Lithium Iron Phosphate battery ($LiFePO_4$)
Li-LTO	Lithium Titanite Oxide battery ($Li_4Ti_5O_{12}$)
DoD	Depth of Discharge of battery
C-rate	Charge and discharge rate compared to batteries nominal energy content value
PEM	Polymer Electrolyte Membrane fuel cell
SOFC	Solid Oxide Fuel Cell
NM	Nautical Mile
MW	Megawatt of power
MWh	Megawatt hour of energy
kW	Kilowatt of power
kWh	Kilowatt hour of energy
kn	Knot

1. INTRODUCTION

Recent developments open the way for ship owners to enhance energy efficiency using battery energy storage, fuel cells and thermal energy storage as part of powering ships, or as the sole source of energy for powering and propelling ships. The alternative a ship owner is selecting, depends mostly on the ship type, operational profile, requirement for safety and redundancy. The trend towards using batteries for energy storage, either to support a power plant or for full battery power began a few years ago. For some ship types, batteries seem to be an obvious choice, while other ship types are struggling to find a solution that fits best into their profile and characteristics.

This paper will touch on some basic elements of current technology, including the limits and possibilities it creates. The paper will also introduce future limitations and possibilities, focusing mostly on battery energy storage and existing fuel cell technology.

Additionally, the definition and possible benefit of thermal energy storage will be discussed.

2. BATTERY ENERGY STORAGE

A power plant for a modern, large cruise ship is between 50 and 70 MW, while a smaller scale commuter ferry carrying passengers and cars has a power range of between 0.5 and 3.0 MW. A typical operational profile for the cruise ship sees it at sea half of its operational time, giving an average sea time of 12h/day. A typical operational profile for the commuter ferry is 5 to 20 min transit between harbours, with harbour time between 4 to 20 min.

What does this tell us about the possibilities and/or challenges of introducing battery energy storage to these types of ships using battery technology as it exists today? How will the same picture look five years from now?

Today, approximately 70MWh of battery energy storage has been delivered, or is being delivered to the shipping industry. Looking at the size of the power plant for a cruise ship, one realizes that the market and business size is still relatively small. 70MWh is enough for powering a large cruise ship for a total of 2.5 hours at a speed of 17kn.

2.1 PURPOSE OF BATTERIES FOR SHIPS

The typical application and use for battery energy storage onboard ships vary a lot between different ship types, but from the passenger ship perspective the most promising areas for use are:

- Full battery operation. Mostly usable for commuter traffic and/or short sea transits. The largest benefits come from emissions- and energy-saving perspectives.
- Spinning reserve. Usable for all ships, where redundancy requirements are high but actual power demand is moderate; power for redundancy can be guaranteed through this type of solution.
- Peak Load Shaving. Supporting combustion engine efficiency, enabling engines to operate on maximum efficiency.

2.2 BATTERY TECHNOLOGY

All ships use batteries today for UPS (uninterrupted power supply) applications, typically Lead Acid batteries. In this paper, all the batteries being discussed are of the Li-Ion type, although the chemistry may differ, in being either Li-NMC, Li-LFP or Li-LTO.

The most commonly used chemistry amongst the batteries is Li-NMC, where the mixture of Nickel, Manganese and Cobalt represent the main characteristics of the battery. Different formulations influence the charge, discharge rates, the number of cycles and the life time of batteries. Batteries used in marine applications have, historically, been specified for high charge and discharge rates, sacrificing expectations on lifespan. A typical charge rate for a battery installed on a commuter ferry is 2C, resulting in relatively big losses due to the heat generated in the cells. To take care of the heat generated, special liquid-cooled batteries have been developed for the market. The efficiency of a liquid-cooled Li-NMC battery is one of the highest on the market.

Looking at trends developing among cell and battery manufacturers, it seems that there is a shift from the high C-rate applications towards a more moderate C-rate application. Why? The biggest reason is increased density, where energy is increasing compared to volume at the same time as the price per kWh is going down. The result of this is simpler batteries, less loss and a longer life expectancy for batteries. In late 2016, one battery maker was offering a 6.5kWh battery module with dimensions of approximately 36 cm by 30 cm by 30 cm. Six months later, the same supplier is offering a 10kWh battery module of the same size, but with lower charge/discharge rates than the battery 6.5kWh module. However, with lower charge/discharge rates, the importance of high C-rates can be minimized with higher density and lower price per unit. With lower C-rates and improved manufacturing process we can expect the number of cycles to increase from today's 6000 cycles to 12000 cycles, and even higher, enabling a longer battery life time.

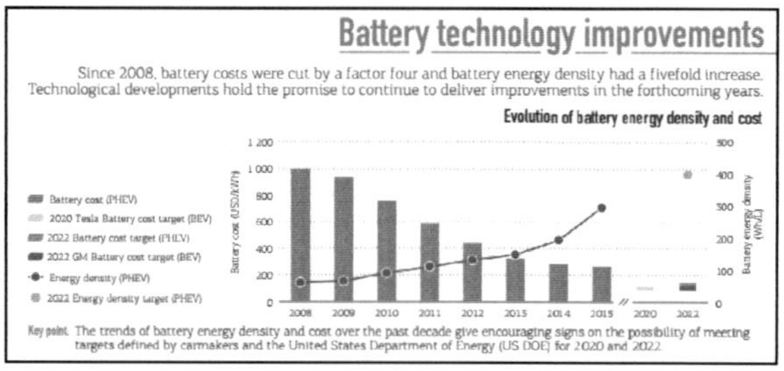

Figure 1. Global Outlook EV 2016, IEA org

Figure 1 gives an overview of the energy density improvements of Li-Ion batteries for electric vehicles, where the density improvement between the years 2014 and 2015 was 33%. The projected improvement to 2022 from 2015 is an additional 25%. Comparing the energy density improvements achieved by battery systems in Electric Vehicles with the recent energy density improvements in Marine-type of batteries, it is likely that similar energy density improvements will be achieved by Marine batteries in the future.

2.3 EXAMPLE CALCULATION FOR BATTERIES

This chapter considers a commuter ferry that has been converted to operate fully on battery power. The speed of the ferry is 10 kn and its travel distance is 3.33NM, which takes 20 min. The propulsion power demand is 1,500 kW with the base load power demand of 150 kW. The parameters for comparison are the weight and energy density of the batteries in four different options;

- battery module with 6.5kWh energy;
- increased energy capacity for battery module to 10kWh;
- 10kWh battery module with improved hull form and higher propulsion efficiency;
- a battery size of 12.5kWh with the same foot print as the 6.5kWh battery module, including improved hull form and higher propulsion efficiency

Graph 1 show a comparison between battery modules with 6.5kWh and 10kWh of energy. The weight difference is significant between these batteries and the corresponding foot print is approximately 30% less for the same operation of the batteries. The weight saving will also have an impact on the energy consumption for the ferry. In the comparison, we have used the same hull form and propulsion efficiency. The third option with 10kWh battery module and improved ship hull form and better propulsion efficiency will reduce the overall energy consumption by an additional 6%. The fourth comparison is for a battery module of 12.5kWh, which could be reality by 2022. The weight reduction is over 50% between the 6.5kWh battery module and 12.5kWh battery module.

Graph 2 uses the same comparison parameters, but is considered from the point of view of the distance the ferry can travel with improved energy density and improved hull form and propulsion efficiency. In this comparison, the weight will be the same for all options.

The Depth of Discharge (DoD) is limited to 60% in the base calculation and one of the dimensional parameters is based on this. Graph 3 indicates the expected DoD if the system was designed to keep the DoD as small as possible, thus maximizing the lifetime of the batteries. The distance travelled is 3.33NM for all options.

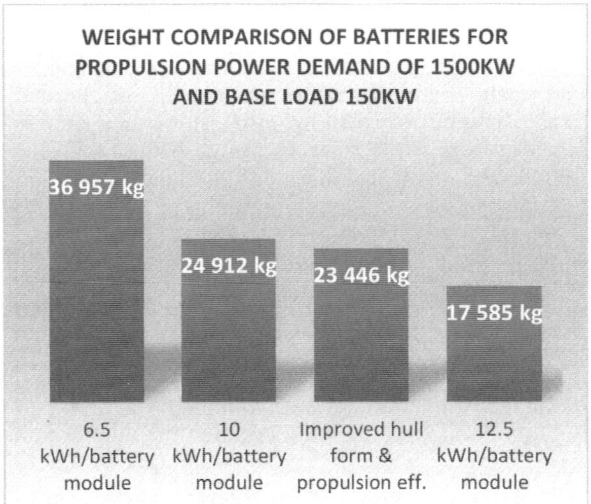

Graph 1. Weight comparison

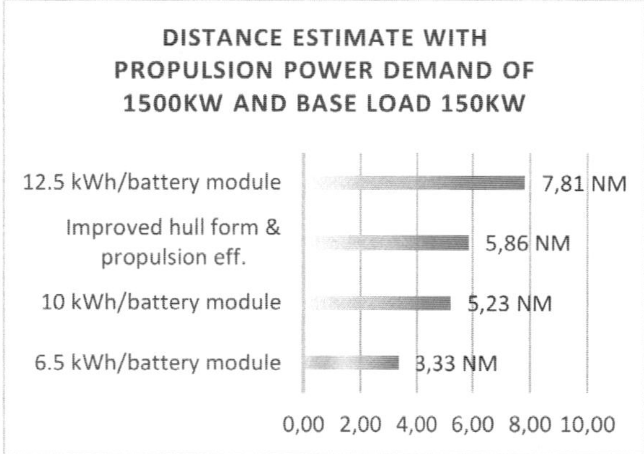

Graph 2. Distance comparison

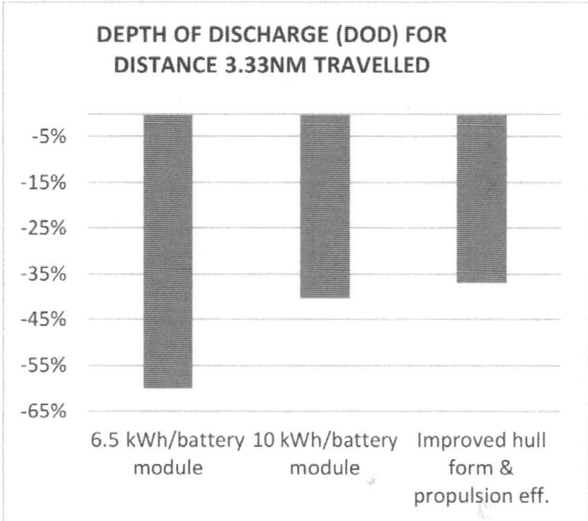

Graph 3. Depth of Discharge for 3.33NM distance travelled.

Even though the example above is calculated for a commuter ferry, a river cruise ship is another ship type that would greatly benefit from full battery operation, either in port or at slow speed cruising. Charging stations should be available at designated areas for this purpose.

2.4 BATTERIES FOR OTHER APPLICATIONS

The most likely battery applications for a cruise ship in the near future is as a spinning reserve and for peak load shaving. In some cases, we will see cruise ship owners seeking to use battery energy storage systems to operate the ship in full battery mode when entering the harbour. This would require a large effort on reducing the base load on the ship, in combination with an energy efficient propulsion system. This may be a realistic scenario by 2022 using a 12.5 kWh battery module if unit weights can be minimised. A 3000kWh system, weighing around

35000 kg, could yield as much as 7200kW over a 15-minute period. The DoD for such a system is 60%, enabling a long enough life time for the investment.

A relatively large benefit can be achieved with comparative ease if batteries are used as spinning reserve, allowing engines to operate at optimal load, even in areas where high redundancy is required. The savings are achieved optimizing the number of engines on-line at all times.

Due to the nature of its operation, an exploration cruise ship fits very well into operating batteries as a spinning reserve. These ships often operate in sensitive areas and in low load conditions, with high redundancy requirements. Using batteries as a spinning reserve will enable better engine efficiency and lower emissions.

The size of the battery to support a spinning reserve at slow speeds is relatively small - between 500 kWh and 720 kWh with the battery operating with a maximum discharge rate of 3C. Considering a maximum DoD of 70%, the power in use is between 1050 kW and 1500 kW for 20 min in a typical exploration cruise ship. This will be more than enough time to start an engine.

3. FUEL CELL

Recently Foreship has also seen an increased interest in using fuel cells in parallel with combustion engines to improve fuel efficiency and reduce emissions. For the time being, the most promising of the various technologies available is PEM – Polymer Electrolyte Membrane, whose use in the automotive industry has brought a lower price per energy unit than the less mature Solid Oxide Fuel Cell (SOFC). Another strong advantage for PEM over SOFC, at least for the time being, is its much better power density. However, SOFC efficiency can be raised up to 65-70%, while PEM has a lower efficiency of around 45%. Therefore, a lot of effort is currently being put into the development of the SOFC technology. For this reason, it is to be expected that within the next ten years SOFC will become a significant competitor to PEM technology. Other fuel cell technologies also exist but their potential is limited in the marine market due to the superior performance of PEM and SOFC technologies.

3.1 FUEL CELL TECHNOLOGY

Both PEM and SOFC technologies use hydrogen as fuel, but other types of fuel can be used as the source for the hydrogen such as LNG and Methanol. However, both LNG and Methanol create emissions and need a reforming process to produce hydrogen, with higher cost and foot print demand resulting. After the reforming process, both carbon monoxide and hydrogen remain, and it is worth noting that while SOFC plants can utilize both carbon monoxide and hydrogen as fuel, the carbon monoxide still needs to be removed in the case of PEM, thus lowering the overall efficiency and complicating the fuel treatment process.

While the SOFC has a clear advantage with respect to system efficiency over PEM, the higher operating temperature (750 °C vs. 160 °C) means that the auxiliary components of the system are more expensive and complicated, with start-up and shut-down cycles significantly longer. This also means that load changes can be achieved much more rapidly with PEM technology.

One important matter that needs attention is the cell (cell stack) lifetime of the different technologies. As the PEM development is mainly driven by the automotive industry, where the overall lifetime is not considered as important as for applications where the system is constantly at high load over long periods of time, the SOFC might already have a clear advantage. This is one of the key elements to be looked at when comparing technology alternatives, as the cells are the most expensive part of the system.

As there are several commercial marine fuel cell pilot installation projects ongoing in the range of hundreds of kilowatts per installation, the belief is that within the next ten years fuel cells will start becoming more common on ships. The next step will be megawatt-size installations, where power density will improve and price per kilowatt goes down. For the time being, though, both technologies have a relatively high cost per energy output.

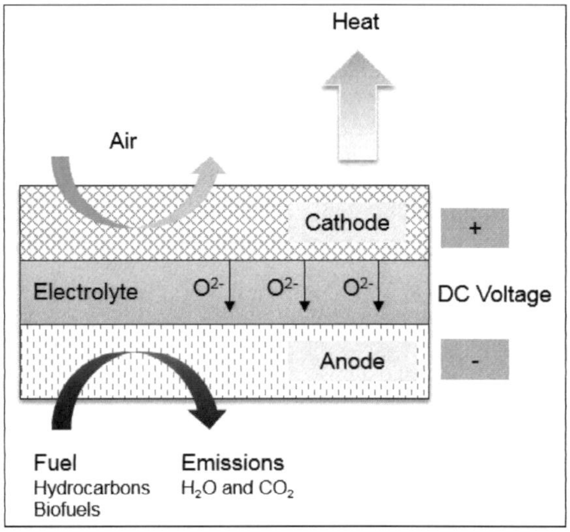

Figure 2. Working principle of a fuel cell

A fuel cell is an electrochemical device, see figure 2, that directly converts the chemical energy of a fuel into electricity and heat. It produces low-voltage DC directly from the fuel and it comprises of three functional blocks:

- cathode: reduction of oxidant

- electrolyte: transfer of ions
- anode: oxidation of fuel

A fuel cell power plant comprises of several cell stacks – stacks of cells piled on top of each other

4. THERMAL ENERGY STORAGE

The big power plants on cruise ships generate large amounts of waste heat. Therefore, utilizing excess heat, or cold, is essential. The introduction of thermal energy storage will enable a cruise ship to store excess heat produced in high transit speeds, instead of dumping the heat. Storing excess heat or cold allows the ship to use this heating or cooling capacity in part load conditions or at low transit speeds. There are different technologies available to store thermal energy: a water tank is the simplest storage solution, but water requires tanks that are large in volume and loses energy if stored for a long time. Different phase changing methods and technologies will enable thermal energy to be stored for longer periods with less demand for volume. Another way of storing excess heat is to increase the HFO storage tank temperature when excess heat is available, to keep the peak heat load demand to a minimum in slow speed transits, or in port conditions.

In the past, the need to store thermal energy has not existed because most of the energy has been used, either in heating HFO tanks and fuels and/or by fresh water production. With the introduction of LNG as a fuel, as well as alternative technologies for fresh water production, the possibility of using this energy for something else has increased.

Thermal energy storage in combination with a well-planned and built waste heat energy recovery system can reduce emissions significantly onboard a large passenger ship.

5. CONCLUSIONS

Battery energy storage is clearly beneficial on short sea routes, and with the expected improved energy density and lower price per kilowatt hour we will see many more commuter ferries either converting to battery energy storage or being built as battery-operated ships from the beginning.

Early cruise adopters of battery technology include ship operators calling in particularly sensitive areas where there are clear environmental incentives to invest in battery technology. However, it can be difficult to move from case studies to generalisations in terms of larger scale uptake for cruise ships. A combination of fuel cells, batteries and combustion engines is the most likely scenario for cruise ships.

It's not likely that we will see large numbers of passenger ships operating solely on batteries, but the belief is that 40-60% of all vessels could benefit from auxiliary battery load exploitation to support peak load shaving. A small number of ships – say, 5% of the fleet – will also use them for specific duties: in the cruise sector, for example, battery power could be useful during port entry, where the environmental gains would be strong.

There is a significant interest in fuel cell technology, but if consideration is limited to PEM technology, with its moderate efficiency of 45%, it is unlikely we will see major investment in this technology soon, due to its high cost and the lack of hydrogen as fuel. SOFC technology, with its higher efficiency of up to 70% seems to be more promising, but here the high operating temperature may restrict the use of this technology onboard ships. Both PEM and SOFC technologies require a stable high load to operate efficiently. For this purpose, the battery energy storage will play a significant role in the use of fuel cell technology as well.

6. AUTHORS BIOGRAPHIES

Jan-Erik Räsänen is Head of New Technologies at Foreship Ltd. He is a well-known ship energy expert, focusing on reducing energy consumption onboard ships, and as such enabling ships to minimize the emissions from their power plants.

WIN WIN WIND SITUATION: COMMERCIAL CASE FOR COMMERCIAL WIND PROPULSION

G Allwright, International Windship Association (IWSA)

SUMMARY

We have seen a wind change in the industry over the last 4-5 years where it comes to shipping efficiency and this has gone hand in hand with a steady growth in commercial wind propulsion projects, matched by a very cautious and gradual change in the perceptions over wind propulsion in the industry. Nonetheless momentum has been building and this paper will place that change into perspective, focusing on the change drivers, the barriers & solutions to that and examples of new build and retrofit projects to watch.

This paper outlines the current state of commercial wind propulsion and how we have reached this point from a technical point of view (sail design, rotors, kites and new hull designs), detailing the financial and policy drivers and the market barriers and how those will likely affect uptake of wind propulsion.

1. INTRODUCTION

The Wind energy sector has taken giant strides in offshore wind farm development, the first offshore wind farms were developed in 2011 and the energy produced cost approximately €300 MW/h, that has exponentially fallen today to as low as €30-40 MW/h (Vattenfall, Oct 2017), and is now cheaper than fossil fuel and nuclear. At Burbo Bank, less than 10 years ago , the first 3.6 MW wind turbine was installed. This year 8.0 MW turbines - the world's largest wind turbines – were installed. In less than a decade the turbines have more than doubled in power output and the cost of manufacture, installation at scale and thus the energy produced has fallen dramatically, and wind propulsion for commercial shipping certainly has the potential to follow a similar development curve.

We must remind ourselves of the challenge ahead, the remaining carbon budget we have to achieve the goal of restricting global temperature rises to 1.5C is very limited, with 5 years of current emissions or slightly longer, depending on whether we adopt a hard stop or soft stop/overshoot and remediation approach outlined by the IPCC. IMO projects a doubling of the world fleet by 2050, and world trade growth will lead to 50-250% growth in emissions if no mitigation steps are taken (IMO 2015). With land based emissions set to decline dramatically in the coming decade, a 'business as usual' approach in shipping will see its share of global emissions grow up to 17% of global CO2 emissions in 2050, if left unregulated. (EU 2015)

The industry is starting to wake up to this challenge, with recently released statements for the ICS and other acknowledging the need for 50% reduction in CO2 emissions by 2050, in line with a 2C temperature rise trajectory model (ICS 2017), but many in the industry recognise the need to move faster and see a commercial opportunity in doing that with a higher ambition 1.5C target, in the knowledge that ships designed and built today will need to operate in a net zero emissions world by the end of their operating lives, if we are to keep within the now ratified Paris Agreement to limit global temperature increases to 1.5C.

As the Shipping in Changing Climates project work has shown, the 1.5C scenario will require ultra-low carbon and zero emissions technology uptake and operations to start aggressively in 2030 and grow to full fleet penetration by 2040 and this will mean large scale scrapping and bankruptcy through forced obsolescence and while a 2C scenario pushes that timeline out further, this will still require drastic and far reaching changes through the industry. (SCC 2017)

The last 12 months have seen some significant movement in the development of wind propulsion for shipping and the conversation has continued to evolve, with the 'how' and the 'when' increasingly coming to the fore. This article will highlight some of those changes, give a brief overview of additional project advances, R&D development and commercial uptake of wind propulsion technologies, taking a look at recent significant report releases on the sector and the further development of the International Windship Association (IWSA) itself.

This period of development has been book-ended by the release of a significant report from the EU at the start of the year and the concluding work from the Shipping in Changing Climates team in September 2017 at the other end. The EU DG Clima commissioned report: *Study on the analysis of market potentials and market barriers for wind propulsion technologies for ships*, was jointly carried out by CE Delft, Tyndall Centre for Climate Change Research, Fraunhofer ISI, and Chalmers University of Technology. (CE Delft 2016/17). The main findings from the Shipping in Changing Climates project were showcased over a number of days at the London International Shipping Week, this being a coalition of universities including UCL, University of Strathclyde, University of Manchester, Newcastle University, and University of Southampton, funded by UK Engineering and Physical Sciences Research Council and advised by industry including Lloyds Register, Rolls Royce, Shell, BMT and MSI

Both third party expert teams have identified wind propulsion as a significant addition to the shipping industry's toolbox in its drive towards significantly lower carbon emissions and the movement towards zero-emissions during the second half of the century when focused on the 2C target. As stated before, the ships that are being designed and built today will need to be operating in a net-zero emissions environment by the end of their operational life and both wind-assist and wind-auxiliary solutions will need to be brought in at scale if we are to reach that goal.

The EU report assesses the barriers and drivers to the development of wind propulsion, but the key focus is on the business case for the uptake of wind propulsion and this is an important step forward, with an independent third party assessing the potential growth and impact of the development of this industry segment, the headline finding reads (with noted caveats):

'Should some wind propulsion technologies for ships reach marketability in 2020, the maximum market potential for bulk carriers, tankers and container vessels is estimated to add up to around 3,700-10,700 installed systems until 2030, including both retrofits and installations on new builds, depending on the bunker fuel price, the speed of the vessels, and the discount rate applied. The use of these wind propulsion systems would then lead to CO2 savings of around 3.5-7.5 Mt CO2 in 2030 and the wind propulsion sector would then be good for around 6,500-8,000 direct and around 8,500-10,000 indirect jobs.' (CE Delft, Nov 2016)

The findings of this report feed into the later discussion of barriers and IWSA activities and projects to tackle those. More recently, we had the final reporting from the Shipping in Changing Climates in September and this multi-university, multi-disciplinary systems research project used 'big data' sources such as the Satellite AIS data, ship level smart data, to which the consortium had unparalleled access to assess the pathways, technologies and fuel choices required for the industry to meet a series of climate models

The focus of this project has been multifaceted, however among the key findings of SCC and specifically presented in papers such as CO2 Emissions from International Shipping: Possible reduction targets and their associated pathways, authored by UMAS, UCL and LR, is the need for wind-assist technologies is clear for the mix of propulsion going forward, although the report also highlights the lack of maturity currently (UMAS 2016)

The last 12 years have seen numerous initiatives develop further through IWSA and these have aligned with many of the recommendations made in the aforementioned EU report.

Perception within the industry and among policy makers of the viability and credibility of the technologies has been a major focus and work continues in this field with the dissemination of both wind-assist and wind-auxiliary research and project updates, however there has been a noticeable disconnect between this increasing volume of quality information and deliberations in the policy field. To quantify this, early this year IWSA reached out to approximately 100 flagging nation delegations in the IMO with a questionnaire and information pack regarding wind propulsion. While the response rate to date has been limited, with only 27 full and partial responses (thus not statistically significant), these responses have given an insight into how much information is getting through to the delegates and where that information is coming from.

The three most important questions from IWSA's perspective were the following:

Question 3: Rate your awareness of wind propulsion technologies (in general)

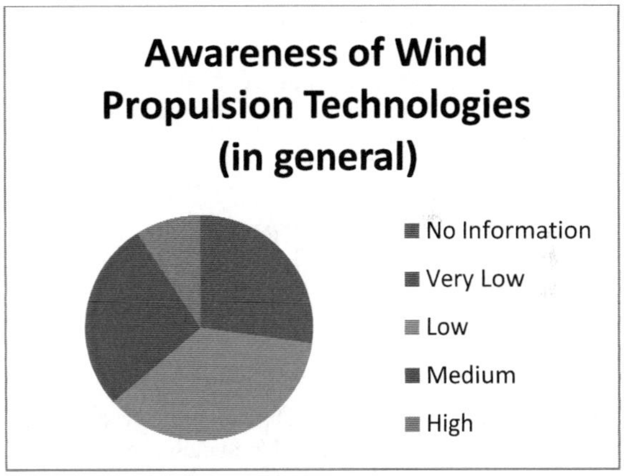

Figure 1 Awareness of wind propulsion

These poor general results were compounded by a clear majority of respondents having little or no knowledge when questioned further about the potential of specific wind propulsion technologies: rotors, suction wings, hard sails etc.

Question 9: How would you rate the availability of information regarding wind propulsion technology?

All respondents rated the availability of information from moderate (33%), down to low/very low.

Question 10: Where does your organisation get its information regarding wind propulsion technology?

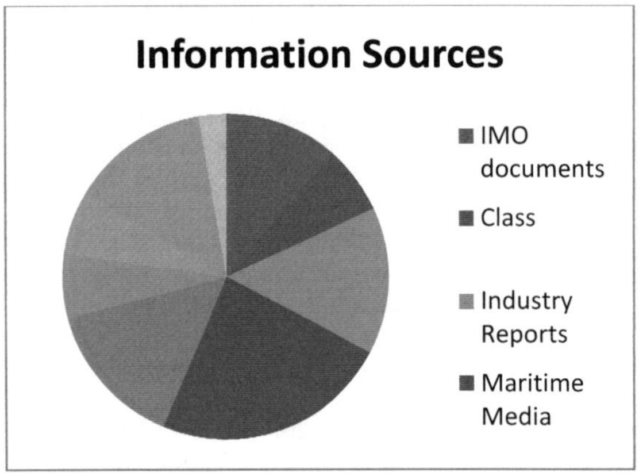

Figure 2 Sources of Information

As we can see for the figure above, the sources of information are quite diverse, however most of the information is being garnered from non-specialist sources, i.e. Media, networks, IMO documents and industry reports, which tend to rely on general perceptions of the technology and projects available rather than in-depth analysis. Particularly worrying, though not unexpected is the lack of information being garnered from specialist work from Class, academia or the IWSA itself and highlights that much more work needs to be done in this field as 90% of respondents indicated that the uptake of wind propulsion technologies was moderate to very high desirability in tackling the decarbonisation challenge.

2. OVERCOMING BARRIERS

Work has also been continuing on other areas highlighted in the EU report, especially concerning the initial crucial steps:

"These key barriers are interrelated in different ways, with the most crucial interaction being a chicken-and-egg problem between the first (Trusted information about wind propulsion technologies) and the second key barrier (Access to capital). In order to breach this chicken-and-egg problem, the report sees the development of a standardized method to assess wind propulsion technologies combined with test cases to develop this assessment method as the most important starting point for overcoming the barriers." (CE Delft 2016/17)

The University of Strathclyde, Glasgow has been working on the development of a 3rd Party Verification tool which will be one such system for helping designers and ship owners to access wind propulsion systems. IWSA is collaborating with the Strathclyde team and our members will be providing technical specifications and where relevant operational data to help with piloting this software early next year. This performance modelling takes more than 30 major design points and simulates performance under varied wind and wave conditions based on decades of weather data. The verification tool enables technology providers to robustly assess their technology and ship projects, and will ultimately enable ship owners to assess wind technology installations across their fleet profiles.

Further work is also being undertaken in the assessment of barriers to the uptake of wind propulsion from a technical and operational perspective. MARIN, an active IWSA member, submitted a study, commissioned by the Netherlands, for consideration at MEPC 70/71: Air Pollution and Energy Efficiency: Study on EEDI reduction factors (MARIN, 2016), which specifically covered wind-assist technologies. This paper focused on the thresholds that ships need to meet are reduced sequentially with the last planned step at 2025. The targets for 2025 are under discussion. The Dutch Ministry of Infrastructure and Environment commissioned MARIN to study if and how much the targets for 2025 may be tightened. The study included general hydrodynamic measures and also included the possibilities offered by wind assisted ship propulsion. Their work in this area is continuing with a joint industry project (JIP) proposal from MARIN and ABS, which is supported by IWSA and others. The Wind-Assisted Ship Propulsion: Proposal for Joint Industry Project objective is set to overcome the barriers to the uptake of wind-assisted propulsion, and specifically to:

- Determine baseline methods for transparent performance prediction, to be proposed as industry standards;
- Use the methods to provide predictions to ship owners/operators for their fleet;
- Review current status of rules and regulations for wind propulsors, identify gaps and make recommendations, and provide examples on establishing compliance. (MARIN & ABS, 2017)

The aim is to cover the majority of all marketed wind-assisted ship propulsors and the project will include technology providers, designers, class, flag states and shipyards, thus covering much of the development chain for this technology uptake.

In addition to this important work, and indeed an extension of the JIP above is the need for a comprehensive approach to wind-assist and wind-auxiliary technology and vessel development.

While specific guidance documents for selected wind-assist technologies have been released, including Guidance Notes for Flettner Rotor Approval (LR 2015) and Certification and Classification Procedures associated with installation of a "Flettner" Rotor Unit on board a classed vessel (DNV/GL 2016), there is a need for a wider approach and work towards a 'Windship Code' will be the final goal, good examples being codes for Large Yachts, High Speed Craft and Special Purpose Ships.

3. COLLABORATION WITHIN THE INDUSTRY

IWSA members have recognised the need for further collaboration within the industry and for a multi-stakeholder approach, if we are to develop a resilient, sustainable and ultimately profitable wind propulsion segment within the industry and two projects are underway that will help to advance that agenda further. Firstly, IWSA is nurturing and supporting the development of 'centres of excellence' or 'wind propulsion hubs' worldwide. These regional hubs are aimed to maximise scarce local resources and create collective cost reductions, co-creation projects and both catalyse and utilise local and regional networks around the development of these technologies and their installations on board vessels.

The front runner over the last couple of years has been the extensive work done in the South Pacific by our members there, primarily by the team led by Dr. Peter Nuttall at University of South Pacific. Tireless work on network building, policy advocacy and research into low carbon and zero emissions sea transport with a strong focus on wind propulsion has galvanised the region and backed the South Pacific nations as the driving force behind the 'High Ambition' coalition at the IMO. This work has also had tangible results, with the start of a multi-million euro project in the Marshall islands: *Transitioning to Low Carbon Sea Transport* which will run until 2022.

The national fleet of the Marshall Islands is dependent on imported fuel and emits a significant amount of greenhouse gases and the project, funded by the German Federal Ministry for the Environment, Nature Conservation, Building and Nuclear Safety (BMUB) aims to reduce the emissions from domestic sea transport. The first part of the project will evaluate the emissions and economic conditions of the national fleet in cooperation with experts from the Micronesian Sustainable Transport Centre. In the second part of the project, a demonstrator vessel will be selected jointly with the Marshall Island Shipping Corporation and equipped with the selected technology package and then sea-trialled.

Additional groundwork has been underway in Europe for the development of a number of hubs. An initial multi-stakeholder meeting was held in June 2017 in Nantes to develop an Atlantic coast hub, with 30+ participants and supported by the Nantes and Pays de Loire regional govt. This area has a number of wind propulsion projects underway, and includes excellent academic, research and testing capacity along with a strong maritime tradition and the industry infrastructure to match. Further meetings are being held in October and November this year to formalise the structure and secure core finance to match the ambition of the project.

As we can see from figure 3, there is substantial work to do worldwide to develop these wind propulsion hubs and preliminary scoping work is now underway in Germany/Netherlands and Scandinavia, with N.E. Asia and North America next on the radar.

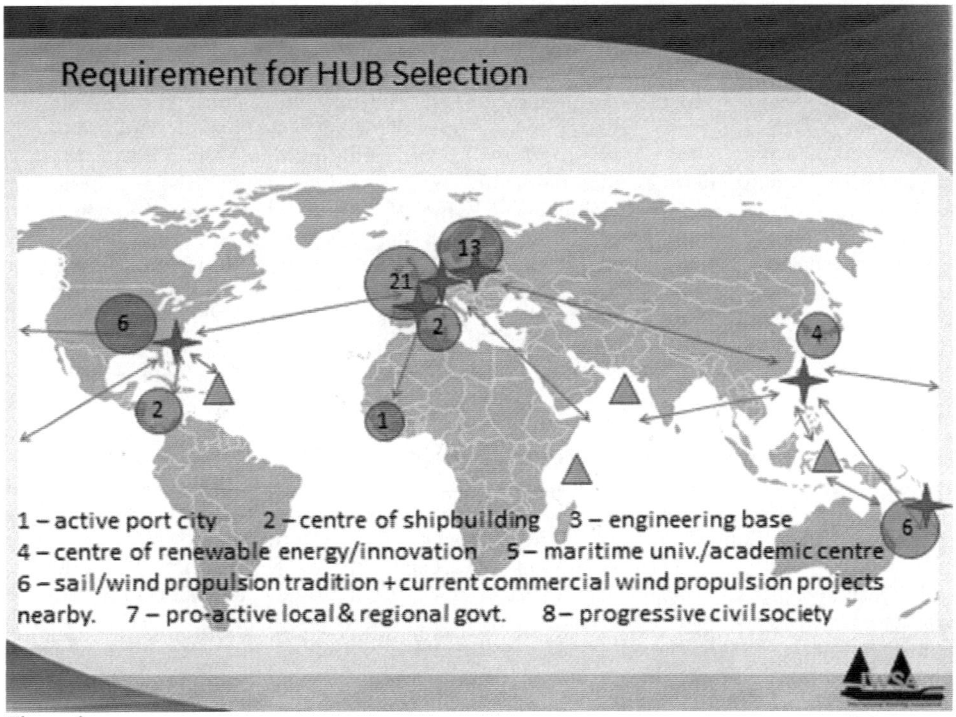

Figure 3

On a smaller scale, but equally important is the IWSA support for the development of the Sail Cargo Network (SCN) in the North Sea region, the team headed up by the Fairwinds Trust www.fairwindstradingcompany.org is a part of the EU Intereg IV B North Sea Region - Dual Port Project. The focus of this project is to analyse and refine business plans and operational profiles for small wind propulsion vessels operating between small entrepreneurial ports in the North Sea region and to create a framework and facilitation tools for existing and new build vessels to trade sustainably in the region, and beyond. This project is also focused on the benefits accruing to both the small ports and the port towns and surrounding region through employment, reduced emissions and increased low carbon trade. www.dualports.eu/sail/

The project lead is the Port of Oostende, Belgium and there are plans to create a dedicated sail cargo quay and renovation of an old customs building to act as a terminal and cargo trans-shipment hub for the trade.

4. WIDER COLLABORATION & NETWORKS

This theme of collaboration and the need to bridge the gap between viability of technology and the capacity and access to capital to bring demonstrators online and to then scale-up has also been driving a number of additional exciting developments. In October, IWSA has entered into a collaborative agreement with the World Wind Energy Association (WWEA) http://www.wwindea.org/, which states on its website:

'The WWEA is an international non-profit association embracing the wind sector worldwide, with more than 600 members in around 100 countries. WWEA works for the promotion and worldwide deployment of wind energy technology.'

Under this agreement IWSA will act as the expert voice in matters of wind propulsion in commercial shipping and WWEA will in turn bring its extensive experience and expertise in the development, bringing to market and scaling-up of innovative low carbon and zero-emissions technologies into the shipping sector. This capacity building will be further enhanced by our exchange of non-executive directors and thus a deeper understanding of each association's challenges and approaches.

A little closer to home, IWSA has also been instrumental in bringing together a group of industry thought-leaders for a day-long event on 13th November at the UNFCCC COP23 proceedings in Bonn. Entitled: *Ambition 1.5C: Global Shipping's Action Plan*, this event will bring together over 150 leading experts in shipping, logistics, decarbonisation and sustainability to start to create an implementation plan to meet the level and speed of decarbonisation required under a strict carbon budget. The aim of the event is to develop an industry-led approach which is sustainable, resilient and profitable – an eye-watering challenge, but potentially mouth-watering opportunity for first movers and early adopters of very low carbon and zero-emissions technologies. The approach is for industry to go out in front of regulation rather than be driven by it and focus on the step-change required in the industry. UNFCCC delegates and NGO's will also be involved and the event is both an official UN Climate partner event and supported by the German Govt. http://www.shippingambition1o5c.com/

The co-organisers include; RightShip, MARIKO, Green Ship of the Future, Lloyd's Register, MAN Diesel & Turbo, Danish Shipping and International Windship Association (IWSA) and they are further supported by the Sustainable Shipping Initiative (SSI), Carbon War Room, UMAS/UCL and the Smart Green Shipping Alliance (SGSA)

5. WIND PROPULSION PROJECT DEVELOPMENTS

As mentioned in earlier sections of this paper, there have been some significant strides forward in project development both in R&D and in market uptake of technology; however the last 12 months has been mainly signified more by the progress that many projects have been making behind the scenes, confidential in many cases due to commercial reasons, than by public announcements. There has been a marked increase in engagement from designers, ship owners, finance and shipyards and the public announcements that have been made, have been important ones, especially in the three categories of wind propulsion system listed below.

5.1 ROTORS

Over the next 12 months, there will be a number of vessels joining Enercon's (Ger) 12,800 dwt prototype E-Ship 1, with her four Flettner rotors which has been in operation worldwide for the last 6 years and the 9,700 dwt MV Estraden, a Bore Shipping RoRo vessel servicing a North Sea route with 2 x 18m Norsepower (Fin) rotorsails fitted since 2015. Maersk and Shell have announced that 2 x 30m rotorsails from Norsepower will be fitted to one of their LR2 110,000-ton product tankers, with an estimated fuel saving of 10% percent. These rotors will be fitted in 2018, with sea trials in early 2019.

Finnish based Viking Lines has also announced the 2018 retrofit of a single 24m version of the same rotors on board the 57,565 GT M/S Viking Grace, their LNG fuelled cruise ferry, which will be projected to save 900 tons of CO_2 per year. This announcement was accompanied by the news that Viking Lines will build and launch its new 63,000 GT vessel in 2020, with additional rotor sails fitted.

(a)

(b)

(c)

Figure 4. (a) E-ship 1; (b) Magnuss Voss and (c) MV Estraden/Norsepower

These vessels will soon be joined the MV Fehn Pollux when she commences trials of the Eco Flettner system. There are a number of other companies working on variants of the standard rotor design, including US-based Magnuss with a retractable telescopic rotor system.

(a)

(b)

(c)

Figure 5. (a) Maersk Tanker + 2 x 30m Rotor Sails ; (b) M/S Viking Grace + 1 x 24m Rotor Sail and (c) Viking Line – New Build

5.2 FIXED SAIL

The Wind Challenger project, headed up by Mitsui O.S.K. Lines and the University of Tokyo, has completed their land-based tests of their telescoping reefed masts. The plan going forward is to retrofit a test rig to one of the MOL vessels next year, which will provide the sea trial data for development of the multi-rigged 84,000 dwt bulker. Another Japanese company Eco Marine Power (EMP) has announced that it has begun preparing for 2018 sea trials of its Aquarius Marine Renewable Energy (MRE) and Energy Sail system in co-operation with the ship owner Hisafuku Kisen K.K. Following that trend, the Japan-based NGO Peace Boat has been moving ahead with its 60,000 GT Ecoship cruise liner project, which features 10 retractable rigid sails, with a LOI signed with the Finnish Arctech Helsinki Shipyard Inc., at Norshipping this year for the build and with the final contract expected shortly. The scheduled delivery date is spring 2020.

Other rigid sail designs are developing quickly, including Windship Technologies (UK) and their Auxiliary Sail Propulsion System (ASPS) which uses fixed wing sail technology. Two automated 35-metre high masts are installed on deck each with three aerodynamic wings fitted. The masts or rigs rotate automatically to exploit the power of the prevailing wind with significant fuel savings modelled. Wind+Wing Technologies (US), Bound4Blue (Esp), Oceanfoil (UK) and Blue

Technology (Den) are also making good progress in R&D or preparations for demonstrator rigs.

Figure 6. a) Peace Boat Ecoship; b) Windship Technology and c) UT Wind Challenger

5.3 AUTOMATED SOFT SAIL

This proven technology continues to attract attention and further development. The Smart Green Shipping Alliance (UK), which includes LR along with Rolls Royce, Tata Steel and others, is continuing development of both its 8000dwt bulker designs and other support systems, with SGSA taking a unique, holistic approach in combining the sail technology, hybrid propulsion, operating systems, operational profile and the circular economy. We have also seen a recent announcement that LR has also joined the Quadriga sustainable shipping project led by Hamburg based Sailing Cargo, which is upscaling the Dutch Ecoliner designs and which also uses the Dynarig system. This project has lengthened and redesigned it to a four-masted, 170-metre car carrier, capable of carrying between 1,700 and 2,000 cars, with a hybrid diesel-electric drive. We can also expect further public announcements soon on the development of soft sail systems, demonstrator vessels and small vessel retrofits and new builds in the near future. The MARIGREEN project initiated by MARIKO and the University of Emden Leer (Ger), is one of those projects that are looking at small, new build soft sail cargo vessels, initially for trading in the North Sea and Baltic regions. The Neoline team in France is also progressing well with their 130-metre, 6,000 dwt RoRo Neoliner designs, which will be primarily powered by sail with a diesel/LNG-electric hybrid auxiliary.

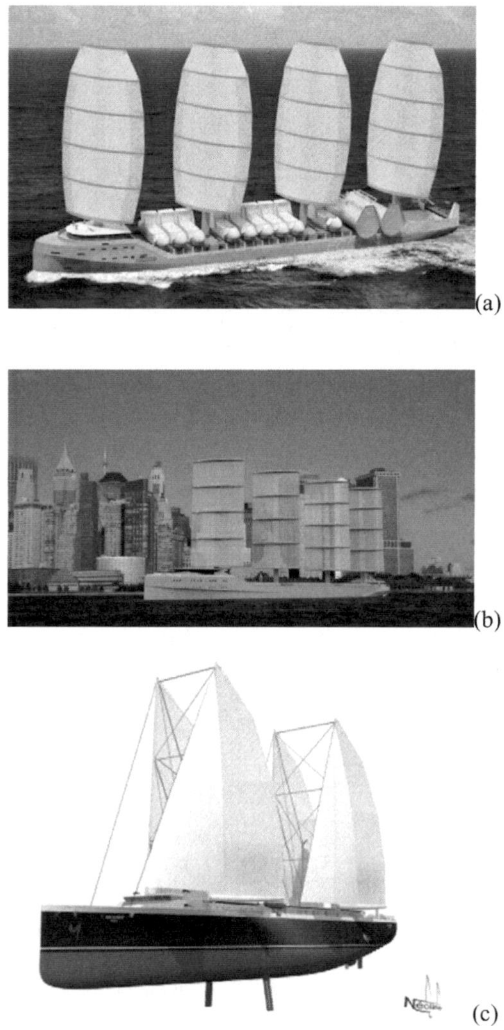

Figure 7. a) Ecoliner; b) Smart Green Shipping Alliance and c) Neoliner

There continues to be progress in other categories of wind propulsion projects and technologies, especially in the development of Suction Wings, Hull configurations and Kites along with Turbines and IWSA looks forward to announcing further strides forward in those technology areas soon.

(a)

(b)

(c)

Figure 8. a) Suction Wing; b) Skysails and c) Vindskip

6. CONCLUSION

We have seen substantial progress within the wind propulsion segment and a lot of foundation work done by the International Windship Association, our members and associates, however there is still much to do to bring wind technology into the main stream of shipping. Nonetheless, with some demonstrators of viable technologies out there in the market, and some large players in the industry backing both retrofit and new build projects, the discussion over wind propulsion has moved on. As an industry, we face a steep and challenging decarbonisation curve, and with a technology basket able to deliver sizeable reductions in emissions and fuel dependency, the discussion has moved from the 'Why' and the 'How' to the 'When' and the 'How Fast'. The first movers and early adopters are positioning themselves to reap the initial benefits of this switch to wind propulsion, the question is when will you jump aboard?

7. RESEARCH UNDERWAY & REFERENCES

1. CE Delft – EU (DG CLIMA) – Analysis of market potentials & market barriers for wind propulsion technologies for ships. (Nov 2016, Public - Jan 2017) http://www.cedelft.eu/publicatie/study_on_the_analysis_of_market_potentials_and_market_barriers_for_wind_propulsion_technologies_for_ships/1891
2. Cleaner Shipping – focus on air pollution, technology and regulation, Danish EcoCouncil (2011) http://ecocouncil.dk/documents/publikationer/1323-130730-shipping-uk/file
3. Dept. of Naval Architecture, Ocean and Marine Engineering, University of Strathclyde - Integrated Wind Assist Ship Performance Prediction(WASPP) and Voyage Level Model (VLM) Analysis Tool http://www.strath.ac.uk/research/subjects/navalarchitectureoceanmarineengineering/sustainableshipping/
4. Emission Reduction Targets for International Aviation & Shipping – Study for ENVI Committee, European Parliament (2015) http://www.europarl.europa.eu/RegData/etudes/STUD/2015/569964/IPOL_STU(2015)569964_EN.pdf
5. EU Intereg IV B North Sea Region - Dual Port Project (2016-2020). www.dualports.eu/sail/
6. International Windship Association (IWSA) – Research centre – listings of many of the publicly available research papers (2017) http://wind-ship.org/en/research/
7. International Windship Association - Work Streams on Commercial Wind Propulsion in Shipping – Market Transformation, Communication, Policy, Technical & Finance www.wind-ship.org
8. International Renewable Energy Agency (IRENA) – Renewable Energy Option in Shipping: Tech Brief (2015) http://www.irena.org/DocumentDownloads/Publications/IRENA_Tech_Brief_RE_for%20Shipping_2015.pdf
9. Lloyd's Register – Wind-powered Shipping: Review of commercial, regulatory & technical factors affecting uptake of wind-assisted propulsion. (2015) http://www.lr.org/en/news-and-insight/publications/windpoweredshipping.aspx
10. Low Carbon Shipping & Shipping in Changing Climates Research Consortium (includes UCL, Univ of Strathclyde, Newcastle, Chalmers, Hull & Plymouth) - www.lowcarbonshipping.co.uk
11. MARIN & ABS (2017) - Wind-Assisted Ship Propulsion: Proposal for Joint Industry Project http://www.marin.nl/web/file?uuid=4e8b652e-2fcf-47f5-986d-

82513a210347&owner=0c55e529-0283-4f06-b873-38b017fd8222

12. Oceania Centre for Sustainable Transport (Univ. of South Pacific) – includes recent development: *Sustainable Sea Transport Solutions for SIDS: Transitioning to Low Carbon Shipping Toolkit* with UNCTAD http://unctad.org/en/Pages/DTL/TTL/Infrastructure-and-Services/SFTF-Toolkit.aspx
13. Revealed preferences for energy efficiency in the shipping markets (Aug 2016) V.Prakash, T.Smith, N. Rehmatulla, J.Mitchell, R.Adland (UCL Energy Institute, Carbon War Room, Norwegian School of Economics) http://shippingefficiency.org/sites/shippingefficiency.org/files/Revealed%20preferences%20for%20energy%20efficiency%20in%20shipping%20markets.pdf
14. Ship Hydromechanics Laboratory at Delft University of Technology - calculating aerodynamic forces and their interaction of several wind-assisted propulsion systems on deck http://www.3me.tudelft.nl/shs/
15. Sustainable Shipping Initiative (Maersk, Cargill, BP Shipping, Carnival, ABS, Wartsila, Unilever etc.) Roadmap Vision 2040 (2016) http://www.ssi2040.org/wp-content/uploads/2016/03/SSI-Roadmap-Simple-web.pdf
16. S@IL Program – Roadmap for Sail Transport - EU Intereg IV (2015) http://www.nsrsail.eu/wp-content/uploads/2015/12/Roadmap-SAIL-Transport-WEB-Bonduelle-WP4.pdf.
17. S@IL Program – Engineering Report – EU Intereg IV (2015) http://www.nsrsail.eu/wp-content/uploads/2015/11/SAIL_FinalReport_Engineering.pdf
18. Third IMO Greenhouse Gas Study 2014, IMO Publications (2015) http://www.imo.org/en/OurWork/Environment/PollutionPrevention/AirPollution/Documents/Third%20Greenhouse%20Gas%20Study/GHG3%20Executive%20Summary%20and%20Report.pdf
19. Vattenfall Presentation – Greenport Congress, Port of Amsterdam, 12 October 2017
20. Wind Technologies: Opportunities & Barriers to a Low Carbon Shipping Industry – Executive Summary – Nishatabbas Rehmatulla, Sophia Parker, Tristan Smith, Victoria Stuglis (UCL Energy Institute & Carbon War Room) (2016) http://carbonwarroom.com/01_cwrebulletins/21shippingwind/RMICWR_Shipping_windpaper_ExecSummary_FINAL.pdf
21. Wind Technologies: Opportunities & Barriers to a Low Carbon Shipping Industry – Full Article - Nishatabbas Rehmatulla, Sophia Parker, Tristan Smith, Victoria Stuglis (UCL Energy Institute & Carbon War Room) Marine Policy (Dec 2015) http://ac.els-cdn.com/S0308597X15003917/1-s2.0-S0308597X15003917-main.pdf?_tid=c4939e98-9d09-11e6-988f-00000aacb35f&acdnat=1477657776_54dc268b5557b84d483b65644b515a6d

Table 1. Projects

Blue Technology (DEN)
www.bluetechnology.dk

Bound4Blue (ESP)
www.bound4blue.com

Crain Technologies (FRA)
www.craintechnologies.com

Ecomarine (JPN)
www.ecomarinepower.com

Ecoliner (NDR)
www.nsrsail.eu

EffShip (SWE)
www.effship,com

Enercon/E-ship1 (GER)
www.ship-efficiency.org/onTEAM/pdf/06-STG_Ship_Efficiency_2013_100913_Paper.pdf

Fair Transport (NDR)
www.fairtransport.eu

Fairwinds Trading Co. (UK)
www.fairwindstradingcompany.org

Inerjy (US)
www.inerjy.com

International Windship Assoc.(IWSA)
www.wind-ship.org

Magnuss (US)
www.magnuss.com

Neoline (FRA)
www.neoline.eu/neoline_en.php

Norsepower (FIN)
www.norsepower.com

Oceanfoil (UK)
www.oceanfoil.com

Ocius (AUS)
www.ocius.com.au

Peace Boat Ecoship Project (JPN)
www.ecoship-pb.com

Propelwind (FRA)
www.propelwind.com

Propit (SWE)
www.propit.se

Sailline Fish (UK)
www.saillinefish.com

Seagate Sail (ITA)
www.seagatesail.com

SkySails (GER)
www.skysails.com

Smart Green Shipping Alliance (UK)
www.smartgreenshippingalliance.com

TankerProa (US)
www.tankerproa.com

Timbercoast (GER)
www.timbercoast.com

Vindskip/Lade AS (NOR)
www.ladeas.no

Wallenius Wilhelmsen S/V Orcelle Project (NOR)
www.2wglobal.com/globalassets/environment/orcelle-green-flagship.pdf

Wind Challenger (JPN)
http://wind.k.u-tokyo.ac.jp/index.html

Wind+Wing Technologies (US)
www.windwingtech.com

Windship Technology (UK)
www.windshiptechnology.com

NORSEPOWER ROTOR SAILS – EFFICIENT AND RELIABLE AUXILIARY WIND PROPULSION FOR SHIPS

J Kuuskoski, Norsepower Oy Ltd, Finland

SUMMARY

This paper discusses the Norsepower Rotor Sail Solution, which is a modernised version of the Flettner rotor – a spinning cylinder that uses the Magnus effect to harness the power of wind to propel a ship. The feasibility of a Rotor Sail installation is dependent on the technical and economic aspects of the ship and its operation. The main technical requirement is that the ship's weather deck has sufficient space for the Rotor Sail installation, and that the locations are exposed to wind from all directions. Conversely, the economic feasibility is mainly dependent on wind conditions in the operating area, the time at sea ratio, and naturally the price of fuel, which determines the financial savings achieved by the reduction of fuel consumption. Various aspects related to ship design and operation with Rotor Sails are discussed and recently announced delivery projects for the Viking Line cruise ferry Grace and Maersk Tankers LR2 tanker are presented.

1. INTRODUCTION

Nearly three years of operational experience onboard Bore's roro-vessel *M/S Estraden* has proven Norsepower's Rotor Sail solution to be an efficient, reliable, and commercially viable auxiliary wind propulsion system. The average fuel savings achieved with two Rotor Sails in the North Sea traffic – since installation in 2015 – has been more than 6%. In other words, that's 400 tonnes per year in reduced fuel, almost $200,000 per year in saved fuel costs, and 1,200 tonnes of less carbon emitted (based upon Rotterdam October 2017 MGO price).

Norsepower was established in 2012 with the ambitious target of developing a modernised version of the Flettner rotor, an invention dating back to the 1920's. The challenge was to design a Rotor Sail solution which is economically justifiable as an investment. Currently the financial incentive to install auxiliary wind propulsion is fuel cost savings, however, reduction of CO_2 and other emissions will likely become another source of financial benefits in the future, as a result of existing and forthcoming environmental regulations.

Design aspects, which were considered to be important requirements for a successful product, were ease of use and low life cycle costs. Extensive tests and practical experience on the *M/S Estraden* have confirmed that Norsepower has succeeded in developing a technically and economically sound Rotor Sail solution. Thrust production performance and reliability of the Rotor Sails are according to expectations. Most importantly, the shipowner and crew are very satisfied with the Rotor Sails. The experience has encouraged Norsepower to develop larger Rotor Sail designs to provide auxiliary wind propulsion for vessels of varying sizes.

2. ROTOR SAIL SOLUTION

The Norsepower Rotor Sail Solution, which can be installed on new vessels or retrofitted on existing ships without off-hire costs, is a modernised version of the Flettner rotor; a spinning cylinder that uses the Magnus effect to harness wind power to propel a ship. The solution is fully automated and senses whenever the wind is strong enough to deliver fuel savings, at which point the rotors start automatically – optimising crew time and resource.

3. ROTOR SAIL OPERATING PRINCIPLE

The operating principle of the Norsepower Rotor Sail Solution – known as the Magnus effect – is described using Figure 1. Figure 2 visualises the wind direction and thrust produced at an angle of 90 degrees to the wind direction. An example of a polar diagram in Figure 3 describes how the thrust (expressed as main-engine equivalent propulsion power) produced by the Rotor Sail changes with the true wind speed and direction. As can be seen on the polar diagram, the Rotor Sails are able to achieve fuel savings from a 20 degrees true wind angle, when the true wind speed is at least 7 m/s. The savings are maximised when the true wind angle is circa 120 degrees and when the true wind speed is 20 m/s or more.

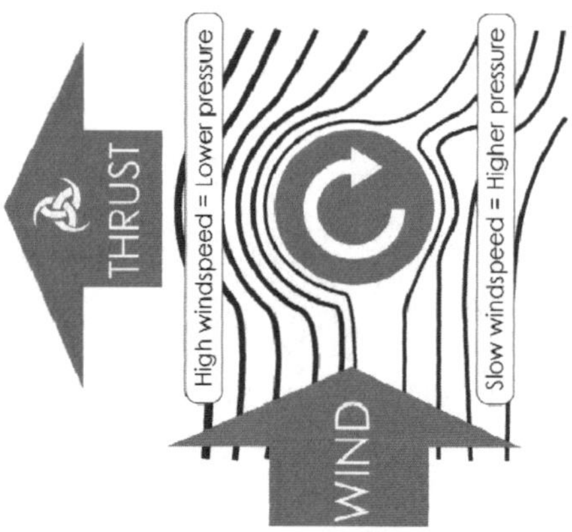

Figure 1. The Magnus effect is caused by a thin layer of air which rotates along the rotor surface and meets the wind on opposite sides of the rotor. On the high (accelerated) wind speed side the air pressure is reduced and on the low (decelerated) wind speed side the air pressure is increased. The pressure difference creates the lift (thrust).

Figure 2. A Rotor Sail produces lift acting at a 90 degree angle from the wind direction. By reversing the direction of rotation the lift force direction changes by 180 degrees.

4. WHAT KINDS OF SHIPS CAN BENEFIT FROM ROTOR SAILS?

Norsepower Rotor Sails are suitable for installation on a variety of ship types and installation is possible for both newbuildings and retrofit projects. Suitable vessel types include ro-ro ships, tankers, bulk carriers, car carriers, ferries, and cruise passenger vessels.

The main technical requirements for installing a Rotor Sail on a vessel are:

- Available deck space in a location appropriately exposed to wind
- Sufficient size of the vessel to carry the weight and maintain the stability requirements

When those two requirements are fulfilled, the focus is usually on evaluating the economic feasibility. Norsepower has developed a simulation software which provides a good basis for evaluating the economic feasibility of a Rotor Sail installation. The following aspects are analysed:

4.1 WIND CONDITIONS ON THE SHIP'S OPERATING ROUTE

The wind conditions determine the potential for delivering sufficient thrust which will translate into fuel savings. The higher the wind speed and the more likely the wind direction is from the beam of the ship, the more thrust a Rotor Sail can produce on average. Norsepower simulates a ship's operation and utilises long term wind statistics from the ship's route area to estimate the propulsion power and fuel consumption savings potential for a specific vessel.

4.2 THE SHIP'S OPERATING PROFILE

A high ratio of time at sea compared to time in port will produce higher savings. In typical cases, the amount of fuel saved increases if the service speed of the ship is increased. Conversely however, the savings percentage also decreases, as the propulsion power requirement increases with a higher rate than the savings from the Rotor Sails.

4.3 INVESTMENT PROFITABILITY

Pay-back period and ROI are calculated for the Rotor Sail investment. A sensitivity analysis of profitability is done with different fuel prices.

For an existing vessel's retrofit project, the steel structure will require some modification to accommodate the foundation block, as well as carry the loads to be transferred from the Rotor Sail to the ship's hull. Cost estimates are case specific and should be done together with the shipowner.

A good estimate of the feasibility of a Rotor Sail installation can be made by carrying out the above evaluations.

4.4 A ROTOR SAIL GENERAL ARRANGEMENT

Norsepower Rotor Sails are available in three sizes with different Rotor Sail heights of 18, 24, and 30 metres and corresponding rotor diameters of 3, 4, and 5 metres.

The Norsepower Rotor Sails are installed on the deck of the vessel with vessel-tailored foundations. For a retrofit project, the foundations and cabling are typically installed during a regular maintenance docking. The rotors are installed on the foundations with a flange and bolt connection. When the installation of the foundations has been completed, the rotors can be lifted onto the vessel and attached to the foundations during a normal harbour stay.

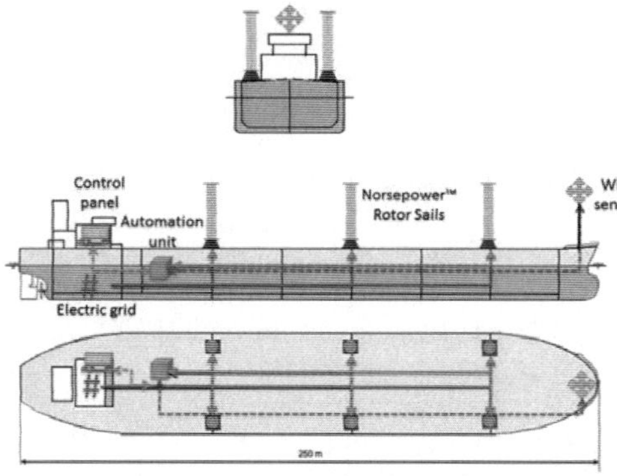

Figure 3.Rotor Sail foundation blocks are welded and integrated to the ship's steel structure. Cabling from the low voltage switchboard provides the required auxiliary power for the Rotor Sails and automation cabling connects the weather station sensors, Rotor Sails, and the control panel located on the bridge.

5. PILOT PROJECT – M/V ESTRADEN

A vital part of developing new technological solutions for marine applications is extensive full-scale testing and operation in the actual operating environment. Norsepower was very fortunate to have Bore as the first shipowner who was prepared to install a Rotor Sail unit onboard their 9,700DWT roro-vessel *M/V Estraden*. The first 18 m tall and 3 m diameter Rotor Sail was installed in the autumn of 2014. Another Rotor Sail unit was installed one year later to further improve the fuel savings and gain more benefit of wind assisted propulsion. The two Rotor Sail units have been thoroughly tested and measurements have been made to verify the performance and integrity of the design.

5.1 EXPERIENCES OF ROTOR SAILS IN OPERATION

After a short commissioning period, the Rotor Sails and the automation system were operating fully, independently, and as designed. The first Rotor Sail was taken directly into operation after installation and commissioning. The first month of operation for the vessel was on a route between Turku, Finland and Harwich, UK.

The installation of the two Rotor Sails was done one at a time with a year of operation in-between to see the possible effect on the ship's normal operations, manoeuvrability, and to confirm the system's reliability and effectiveness. It was evident relatively quickly that there was no noticeable effect on rudder angles or leeway. The system was working reliably and did not cause excess vibrations or noise. The crew also noted that the system had a stabilising effect on the roll motion of the vessel.

Some adjustments were made based on the early experiences, for example, the original location of the wind sensors had to be changed. Even though the sensors were quite far from the Rotor Sails, the down wash of the operating Rotor Sail turned the air flow direction by 90 degrees in some cases. (Väinämö, 2016)

During the first year of operation, the Rotor Sail was unavailable for operation for only a few days only due to a minor mechanical issue and a communication error. During the first month of operation, December 2014, the system was also operated in winter conditions with sub-zero temperatures and events of snowing. No ice build-up on the rotor surface was observed. There is a possibility of ice build-up when the rotors are running on low rpm in straight head wind or idling in harbour. The risk of ice build-up on the rotor surface can be minimised with specific operational modes and other icing prevention measures, for example coatings.

5.2 ACHIEVED FUEL SAVINGS

The average fuel savings achieved with two Rotor Sails in the North Sea traffic has been more than 6% – equating to circa 400 tonnes less fuel consumed and roughly $200,000 per year in reduced fuel costs (based on Rotterdam October 2017 MGO price). This saving has been measured and independently verified by NAPA, the leading maritime data analysis, software and services provider.

The verification of fuel savings and operational performance of the two 18x3 Rotor Sails have given Norsepower confidence to proceed with scaling up the Rotor Sail designs. Two larger models, the 24x4 (24 m high and 4 m diameter) and 30x5 (30 m high and 5 m diameter) Rotor Sails have been launched to achieve even larger savings. The first units of the larger models will also be thoroughly tested on a land test site before delivery to the ship. The first deliveries of both new models will take place in 2018.

Figure 4. Roro vessel M/V Estraden – the first vessel equipped with Norsepower Rotor Sails.

6. VIKING GRACE

The 57,565 GT *M/S Viking Grace* currently operates in the archipelago between Turku (Finland) and Stockholm (Sweden), and is already one of the most environmentally-friendly cruise ferries in the global maritime industry. With the addition of Norsepower's technology, the vessel will further reduce its emissions, fuel burn, and fuel costs; reducing carbon emissions by circa 900 tonnes annually; equivalent to cutting 300 tonnes of LNG fuel per year.

Preparations for the retrofit are underway, with the installation scheduled to take place during Q2 of 2018. *Viking Grace* is set to be retrofitted with one medium-sized 24x4 Norsepower Rotor Sail unit that is 24 m in height and 4 m in diameter.

Figure 5. Illustration of Viking Grace with one 24x4 Norsepower Rotor Sail

7. MAERSK P-CLASS TANKER

In a project co-funded by a consortium including Maersk, Shell, and the UK's Energy Technologies Institute (ETI), a Maersk P-class 109,647 deadweight tonne (DWT) oil products tanker, will be retrofitted with two Norsepower Rotor Sails, which are 30 m high and 5 m in diameter. Together, they are projected to reduce the average fuel consumption on typical global shipping routes by 7-10%.

This project is an important break-through for the tanker market which is considered to be one of the major markets where Rotor Sails provide a competitive solution for improving the fuel efficiency and reducing exhaust gas emissions.

Figure 6. Maersk P-Class illustration with two 30x5 Norsepower Rotor Sails

8. FUTURE VISION

Norsepower sees an excellent potential for the growth of the Rotor Sail market. The company has been extremely satisfied with the operational performance and fuel saving results experienced to date. Going forward, Norsepower plans to ensure continuous optimisation of its technology to further improve the performance and competitiveness of its Rotor Sail technology, and to deliver even greater fuel cost savings.

The long term fuel price predictions vary depending on scenarios, however it can be expected that fuel prices are more likely to increase than to decrease due to the sulphur regulations coming into force in 2020. Norsepower's technology will also reduce the risk of any potential cost increases related to GHG emissions.

9. CONCLUSION

Norsepower is proud to be the first company to utilise the renewable energy of wind for propulsion on a large commercial vessel. As Norsepower continues to grow, we will be able to develop larger and more cost optimised Rotor Sails, which will equate to more fuel savings and improved investment profitability for shipowners and cargo owners paying for the fuel.

The ability to harness the wind as an additional power source to enable a reduction in fuel consumption is a natural next step for the maritime transport industry as it seeks to remain cost-efficient and meet environmental regulations.

11. REFERENCES

1. VÄINÄMÖ J., Operational experiences and results from the first reference installation from Nov-2014, RoRo-ship Estraden (9700 DWT), 24th International HISWA Symposium on Yacht Design and Yacht Construction, Nov. 14-15, 2016.

10. AUTHOR BIOGRAPHY

Jukka Kuuskoski has a M.Sc. (Naval Architecture and Marine Engineering) and an eMBA from the Helsinki University of Technology. He worked with ABB's marine business unit for 15 years in management positions of delivery projects, sales and marketing and service operations. During the past 7 years, Jukka worked in the wind power industry with Winwind Oy, a wind turbine manufacturer, and with Element Power, a wind power project developer. He joined Norsepower in 2016 and is responsible for the sales and marketing of Norsepower's Rotor Sails.

INFLUENCE ANALYSIS OF THE WINGSAIL SYSTEM CONFIGURATION ON THE TECHNICAL PERFORMANCES AND FINANCIAL RETURN OF THE TECHNOLOGY ON SAIL-ASSISTED VESSELS

D Ferrer Desclaux, **M Mantilla Sánchez**, **A García Plaza**, **J M Bermúdez Miquel**, **C Aleixendri Muñoz**, Research and Development Department, Bound 4 Blue S.L., Spain

SUMMARY

Shipping industry's economic viability is highly influenced by fuel-related OPEX, being at the mercy of fuel volatile cost, with an increased pressure from IMO regulations. Various fuel efficiency technologies exist to face this challenge, being wind-based technologies the ones having the potential for double-digit fuel savings. bound4blue is a wind-based fuel efficiency technology supplier, having developed a light weight rigid wingsail technology. The goal of this research paper is to analyse, for a certain vessel, route and rigid wingsail technology, how the rigid wingsail configuration (wingsail size and number) modifies the expected sail-assisted vessel performances and economic parameters, and how these impacts meet or not the maritime transport financial, operational and technical requirements. For such investigation, bound4blue has employed a software tool that determines the performance and financial parameters of a certain sail-assisted vessel in a route. The tool is a combination of 3 modules: a Velocity Prediction Program (VPP), a route and historical weather analysis software, and an economic analysis module. Moreover, a wingsail sizing parametric tool has been used to calculate the expected weight and cost of the wingsail according to its dimensions, aerofoil used and structural parameters, among others. The investigation has been performed for two example case study vessels: a handysize chemical tanker and a car carrier vessel. The research shows that, for the same wind-based technology, the selection of the appropriate configuration is a key factor to meet industry requirements as some scenarios show large variations on the performances and economic parameters.

ACRONYMS

CAPEX	Capital Expense
ECA	Emissions Control Area
EEDI	Energy Efficiency Design Index
HFO	Heavy Fuel Oil
IMO	International Maritime Organisation
MARPOL	International Convention for the Prevention of Pollution from Ships
MGO	Marine Gasoil
NPV	Net Present Value
OPEX	Operative Expense
VPP	Velocity Prediction Program

1. INTRODUCTION

The approval of new regulations is limiting the amount of pollutant emissions allowed to merchant vessels. These new regulations are forcing fuel consumption reduction initiatives and/or the use of low CO_2, SO_X and NO_X emissions fuels, such as the Marine Gasoil (MGO), which has a cost from 50% to 150% higher than Heavy Fuel Oil (HFO) (Ship & Bunker Team, 2017).

The more restrictive regulation framework concerns mainly the sulphur emissions (SO_X). These regulations are limiting the maximum sulphur content of the fuel. Initially, this value was fixed in 4,5% (common HFO) but in 2012 the International Maritime Organisation (IMO) reduced it down to 3,5% (current HFO) (IMO, 2017a).

IMO has set a global limit for sulphur in fuel oil used on board ships of 0,50% m/m (mass by mass) from 1st January 2020 (IMO, 2017a), forcing to use higher-quality and higher-cost fuels meeting this limit, such as MGO. This will significantly reduce the amount of sulphur oxide emanating from ships and should have major health and environmental benefits for the world, particularly for populations living close to ports and coasts (IMO, 2017b), although having an enormous impact on shipping's Operative Expenses (OPEX) (IMO, 2015).

Also, one of the actions taken by the IMO is the definition of the Energy Efficiency Design Index (EEDI), which entered into force in 2013 (IMO, 2011), that aims setting an upper limit to the acceptable pollutant emissions on the new merchant vessels depending on the amount of cargo being transported. Additionally, the European Union has established a set of goals to reduce the environmental impact of transport in Europe, which shows that the strategy of policy makers is to support the appearance of these regulations and more restrictions in the future. The most relevant points include a reduction of 40% (with respect to 2005) of the maritime transport pollutant emissions (European Commission, 2017).

Unlike SO_X emissions, which are a product of fuel sulphur content, NO_X emissions are created during the combustion process (IMO, 2008). How much NO_X is created, again, depends on the fuel being used, impacting shipping fuel-related OPEX. Ships built after 1st January 2016, will need to meet the more stringent Tier III NOx emission limits inside Emissions Control Areas (ECAs) designated to control NO_X emissions (ECA-NO_X) (IMO, 2017b).

In this context, bound4blue has built a rigid wingsail, more similar to an aeroplane wing than to a conventional sail, but adapted to meet the merchant ships requirements. Thanks to the wind, the wingsails generate part of the thrust required, thus reducing the fuel consumption and thereby the pollutant emissions, all of this with no operational impact on the vessel and/or crew members.

bound4blue's designed and patented rigid wingsail system appears as a new solution because of the application of aeronautical techniques to maritime industry. One of its main features is the complete foldability to the ship's deck (including also the mast) in order to eliminate the thrust and/or parasitic side force, ensuring safety during harsh weather conditions and avoiding interference in port operations. The system is designed to fold automatically in case of power failure. For instance, a wingsail of 20 metres' height and 6.6 metres of chord occupies only 1,5 metres above the ship's deck when folded.

In addition, bound4blue's wingsails can rotate on the vertical axis in order to orientate the leading edge to the appropriate direction, which allows maximising the thrust produced by the wingsails independently of the wind direction. Besides, the system is completely autonomous, so no additional crew or training is required to operate it. However, for safety issues, a control panel is available in the crew's cabin in order to operate the system manually if necessary.

2. TECHNOLOGY TRANSFER

In the History of Wind Assisted Propulsion, although there were some previous examples of sailing ships during the first decades of the 20th century, it was after the 1973 oil crisis when the commercial sailing ships started to be considered a feasible option. This context fostered important developments during the late 1970s and 1980s, ranging from detailed aerodynamic and economical studies to prototype design and testing like those shown in (Kazuyuki *et al*, 2017) and (Palmer, Gifford & Partners, 1985).

Nevertheless, the dawn of the sail-assisted era was still far from reality and the investigations in this field continued being led by academia rather than by shipping companies. This threat to the deployment of the technology was reasonably foreseen by C.T. Nance (Nance, 1985), who advised that the variety of wind capture devices, none of which showed clear superiority, together with the stabilisation of fuel prices in 1983, could hamper the technology transfer.

After a 20-years hiatus, several wind-propulsion initiatives have been undertaken and are now being developed. Despite the lack of significant milestones in the Wind Assisted Propulsion field during the late 20th century, the situation now has changed and the key factors mentioned in *Introduction* have brought the technology into the spotlight once more.

The shipping industry is now in a crucial turning point regarding the considerable amount of fuel consumption its entails. This issue has two important strands, the costs associated to this fuel consumption and the environmental implications of it. Therefore, a great effort from all the stakeholders and organisations involved is taking place in order to catch up with the state-of-the-art solutions that are emerging, as well as with the environmental responsibility that is expected from a modern society.

Thus, one of the key factors that could determine the success of the solutions proposed is the proper technology transfer from the R&D companies and universities to the shipping industry. For that to happen, the pivotal idea is profitability. In this sense, the major challenge to face by researchers is to finally make available cost-effective solutions to shipping companies that will pave the way for more extensive research and technology refinement.

From the 1980s investigations to the nowadays literature about the topic, very few considerations about the economic profitability of the analysed systems and technologies can be found. Most of the studies are focused on the absolute performance of the system, usually measured as fuel savings (in percentage), regardless the cost of the system. Another matter of significance is how important are fuel-related costs of the shipping company, since large fuel savings may get diluted and have a marginal impact if these costs are low whereas, if they are high, a small fuel reduction could entail significant amounts of money.

Moreover, when assessing different types of technologies, the most widely used yardstick to compare them is the fuel savings (again as percentage) that they offer, without taking into consideration their suitability for a certain type of vessel, route or operating conditions. Furthermore, even when analysing a single solution, most of them have no fixed set-up but admit variations of some of their parameters, which must be optimised for each case before comparing them. These factors have a great importance for the system profitability and, consequently, for the technology transfer.

Taking all of this into consideration, the purpose of this paper is to show some of the progress made in this regard within the bound4blue project. Additionally, the ensuing methods and tools will be applied to two different vessels, illustrating the significant impact of the device configuration on the overall performance and economic figures.

3. DEVICE ASSESSMENT TOOL

In order to systematise the calculation of the relevant parameters related to the performance and economics of a certain sail-vessel combination, a modular software was developed. It was identified that, in order to assess the best configuration of the rigid wingsails' system, the process of characterising the sail-vessel combination at the preliminary design stage could be divided into 3 different parts with relation to the physical/economical calculations.

The first of these three modules is the Velocity Prediction Program (VPP), which correspond to the isolated performance of the sail-vessel system under a certain weather condition. The second one is the Route Simulation module, which uses the VPP to obtain the mean performance of the sail-vessel system over a specific route and period - for instance 10 years. The third one is the Economic Analysis module, that recursively uses the Route Simulation module to find the resulting economic and financial parameters of the sail-vessel system.

In addition to these three modules, the process was also divided according to the aim of the requested information. Thus, each of the mentioned modules can be run in three different modes. The first one allows the user to obtain the information related to one specific device configuration. The second one, is intended to show the variation of this information with some of the device configuration parameters. Lastly, the third one can optimise the device configuration according to one objective parameter, which is set previously.

3.1 VELOCITY PREDICTION PROGRAM

It was Justin E. Kerwin (Kerwin, 1975) who first introduced the concept of Velocity Prediction Program in 1976. The main idea of it was to develop a procedure for predicting the velocity of racing yachts. Since then, VPPs have become an essential tool in yachts design (Larsson, 1990). In addition, with the resurgence of the Wind-Assisted Propulsion, these kinds of models/algorithms have found another field of application.

Since the objective of these programs is to solve all or some of the six ship-motion equations, a model of the acting forces and moments is needed. The reason why VPPs have attracted such interest is that the forces and moments that play a role in these equations are tightly coupled. Thus, the resolution process must be carefully studied in order to ensure the convergence without a significant loss of efficiency.

In the developed VPP, regarding the forces modelling, the aerodynamic forces of the wingsails' system are introduced as a function of the wingsails' configuration (aspect ratio, number of sails, span, position…) and can consider 3D effects as well as aerodynamic interference between sails (Lee *et al*, 2016). With respect to the hydrodynamic forces, the hull resistance is introduced as a function of the Froude number. Besides, a maximum list angle limitation is imposed within the roll equation to ensure operative comfort and seakeeping.

For the resolution of the equations, it is considered the possibility that the total propulsive force (aerodynamic, made by the wingsails' system and hydrodynamic, made by the propeller) exceeds the hydrodynamic resistance and, consequently, the ship velocity is increased. This situation is achieved when the engine power cannot be sufficiently reduced to keep constant the total propulsive force when the aerodynamic force increases. An extreme case of that occurs when only the aerodynamic force generated equals the hydrodynamic resistance at the desired ship velocity. However, there might be a lower minimum power limit for the engine to ensure its normal operation making this case more likely to happen.

To summarise, the Velocity Prediction Program allows the user to compute for a single weather condition (or a set of them) the resulting aerodynamic forces made by the wingsails' system with the automatically selected optimum sail trimming. Additionally, the list angle caused by the side force is given. Finally, the power requested to the engine to keep the ship velocity constant, and its corresponding fuel consumption are also calculated. As mentioned before, these parameters can be obtained for a single input data, a range of different configurations or for an optimum condition that is found according to one specific criterion previously defined.

3.2 ROUTE SIMULATION

Wind and the rest of weather phenomena affecting the performances of both the isolated vessel and the sail-vessel combination are highly dependent on the geographical area as well as the time of the year. Due to the great impact of weather on vessels' efficiency and comfort, the study of its changes in time and space is very useful for the marine industry. This issue is addressed by Weather Routing, which tries to find the best way to travel from one port to another following some specific criteria. This art has an ancient history: from the seamen's oral tradition in the fifteenth century to the nowadays sophisticated weather forecasting models.

Thus, taking into account the intrinsic dependence of the wingsails' performance on the wind, it is necessary to assess the performance and economic parameters of the sail-vessel system over the typical route where the ship will operate. In fact, even throughout a route, the wind patterns can change drastically from one point to another and a joint study of the entire route could deteriorate the results obtained. An example of that can be seen in Figure 1, which shows a graphic representation of the wind probability matrix obtained for a section of the route Rotterdam-New Orleans which is near the Rotterdam port. In it, it can be observed how the majority of the winds point to the north-west direction (about a 300° heading measured clockwise from north).

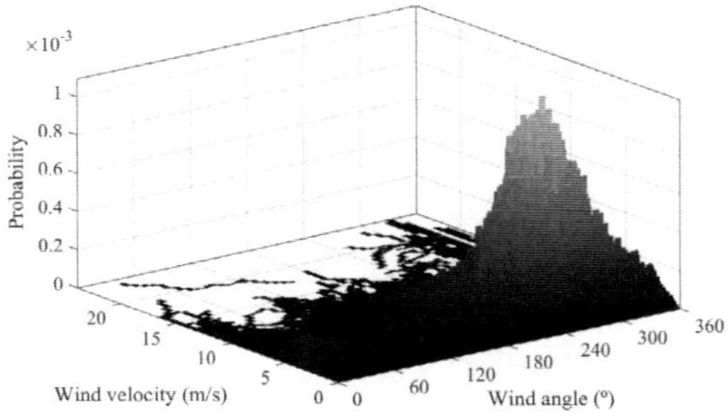

Figure 1: Section 1 of route Rotterdam-New Orleans.

However, in Figure 2, where the same wind probability matrix is plotted for one of the last sections of the same route, that is, near the New Orleans port, the winds are much more distributed. In addition, it is found that there are slightly more winds pointing to East (about a 90° heading measured clockwise from north).

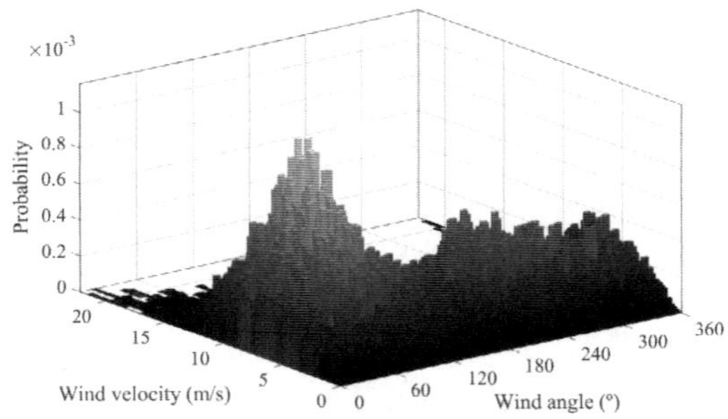

Figure 2: Section 2 of route Rotterdam-New Orleans.

Considering this wind variability, it was decided to create a Route Simulation module that should serve a double purpose:

- Study the weather conditions through a specific route and in a determined period. With this data, the wind probability matrix can be obtained segregated in different route sections. Figures 1 and 2 are two examples of the data obtained with the Route Simulation module.
- Use the obtained segregated wind probability matrix together with the VPP to obtain an overall set of performance parameters of the whole route.

It should be mentioned that the segregation of the route in different sections makes it possible to consider whether each section belongs or not to the so-called Emissions Control Areas. These areas, that were defined by Annex VI of the 1997 MARPOL Protocol and came into effect in 2005 (IMO, 2017b), establish tight regulations about shipping admissible emissions and, consequently, the fuel that should be used within them (Marine Fuel Oil, Heavy Fuel Oil, etc.). Letting the software know when the ship passes through any ECA allows for variations in the fuel used, with the economic effects that it entails.

The option of calculating the route mean parameters after being processed the weather data is not the only way to obtaining a prediction of the sail-vessel behaviour over long periods of time. Another option that was included in the software was perform the simulation of a ship voyage with the actual winds at each position and time from the origin port to the destination and return, repeating from one starting date to some distant date. This second method is more complex and, although offers more control over the simulation parameters, needs more resources. Hence, the selection of one of these two methods will depend on the specific case and the required data.

In addition, the equivalence of the methods was verified comparing the two models within the same periods of time. The result was that the whole simulation parameters of the first method approach the same values obtained with the second method as the periods get longer, so they can be assumed as equivalent.

To conclude, the Route Simulation module creates an interface between the requested averaged route parameters and the wind data (pre-processing it whenever necessary to obtain a series of zonal wind probability matrixes) in order to get a mean value of the ship's performance parameters along the route. Besides, during this process, the fuel savings of the whole route can be calculated. As for the VPP, these parameters can be computed in the three modes: for one single configuration, for several configurations and for an optimal configuration.

3.3 ECONOMIC ANALYSIS

The third module of the Device Assessment Tool is formed by a set of instruments that combine the financial data of the wingsails' system and the vessel, the operation conditions of the vessel and market fuel prices. Moreover, it is able to take into account the effect of changes in the transport capacity as a consequence of the use of the wingsails' system.

Among the financial data related to the installation and operation of the wingsails' system, it is considered the effect of the costs of design and fabrication of the device, the costs of integration on the ship and the costs derived from its maintenance requirements. For the data treatment within the program, these costs are split into Capital Expense (CAPEX) and OPEX, since the former are fixed and do not depend on the operation of the vessel while the latter are a function of the operating conditions. Also, the fuel-related costs of the regular operation of the ship without sails are used to calculate the final economic profitability.

In relation with the operating conditions (including the usual routes where the ship operates), they exert their influence in two different ways. On the one hand, they affect to the calculation of the usual OPEX such as maintenance costs. On the other hand, since the operating conditions substantially modify the performances of the wingsails' system and the fuel savings, they will influence how much the fuel-related OPEX are reduced. This means, the costs reduction offered by the device will change depending on how the ship is operated.

The impact of the device on the system's economics is not restricted to fuel savings but has also an important side effect. Since the velocity might be increased as explained in *Velocity Prediction Program*, the journey duration could be reduced and thus the vessel will have an extra idle time. This time can be used in two different ways:

- Waiting at port until the next scheduled journey. This will reduce the fuel-related costs not only by the reduction in the mean engine power demanded in operation but also by decreasing the engine working time. In other words, the ship transports the same tonnage per year but using the engine less time.
- Increasing the journeys per year of the ship. This will increase the transport capacity and thus the incomes of the shipping company. Nevertheless, the engine will be working the same time annually and fuel costs will reduce slightly less than in the previous case.

The incorporation of this effect into the economic model of the ship operation and how it is done will depend on the operation conditions of the vessel and the characteristics of the cargo transported. Although this issue might be difficult to model, in some cases could significantly increase the profitability of a wingsails' system.

Finally, as mentioned in *Route Simulation*, the Economic Analysis module considers the different prices of the fuels used both inside and outside ECAs. In some cases, the fuel used inside ECAs may cost twice the outside fuel, thus this distinction could significantly change the obtained parameters.

The Economic Analysis module uses all the information mentioned above to obtain some interesting financial parameters that are useful for shipping companies and investment managers in order to decide whether a device suits their necessities or not.

Some of these calculated indicators are the Return of Investment, the Internal Rate of Return or the Net Present Value. As for the two other modules, these calculations might be performed in three different modes. First, for a single configuration, to obtain the actual parameters of a sail-vessel combination. Second, for a range of different configurations, which allows the user to study and understand the behaviour of these parameters with some of the system variables. Last, these parameters can be optimised, obtaining the best configuration in terms of economic profitability. Using this third option the researcher can focus the design process on the final economic profitability for the shipping company, drawing near the technology to the market.

4. PARAMETRIC SIZING TOOL

One of the main characteristics that determine the magnitude of the forces produced by the wingsails' system is its size, which is usually represented by the height of the sails. It is known from the most basic fluid mechanics principles that the aerodynamic force produced by an object within a flow is proportional to its surface, hence the importance of the sails' size for the system performance.

Additionally, since weights and stability are usually critical issues for ships, the weight of the wingsails' system also play a decisive role in the system performance. This weight will depend, among others, on the sail height. Moreover, these two characteristics (sails' height and weight) are at the same time the key factors that determine the final cost of the system, since increasing them inevitably entails larger amounts of material.

Taking this into consideration, it becomes clear that when calculating performances and/or economic parameters (or optimising them) for different sails' height configurations the corresponding weight and cost must be updated. For this purpose, a Parametric Sizing Tool was developed, systematising the calculation of weight and cost for each sail height.

4.1 WINGSAIL STRUCTURE

bound4blue's wingsail design is based on a modular configuration. The structural loads are transmitted through a telescopic mast that allows the wingsail to get folded.

At each stage of the mast it is located a rib. These ribs are the ones that give to the wingsail its aerodynamic shape, since they are manufactured with an aerofoil shape. The selection of the aerofoil is one of the most critical aspects due to its implications for both the structural and aerodynamic properties. The bound4blue's

wingsail can incorporate into the ribs asymmetric aerofoils for high-lift applications using its patented designs. This asymmetry can be achieved both by using cambered aerofoils and by adding some trailing-edge high-lift device such as flaps. Since these ribs are firmly joint to the spar, they all turn together and the rotation is controlled from the mast root. Between two consecutive ribs, a skin based on bound4blue's patented designs covers the structure, serving as a lifting surface.
In Figure 3, it is depicted a scheme of the structural layout of the bound4blue's wingsail design with all the mentioned parts and their arrangement.

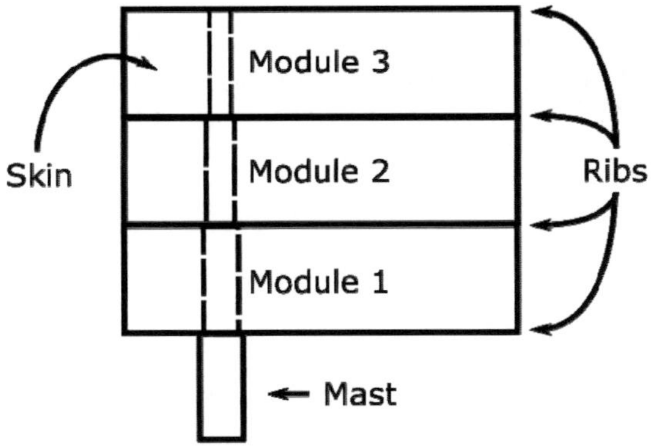

Figure 3: Wingsail structure layout.

4.2 WEIGHT AND COST CALCULATION

For the weight and cost calculation, and accordingly to the wingsail design, there are several parameters that affect the calculation, such as the height of the wingsail, the aerofoil employed, the aspect ratio – i.e. height to width ratio – the number of modules, the maximum load attainable and the properties of the materials used.

Once these parameters have been defined, the tool computes the needed cross-section, thicknesses and other structural dimensions to minimise the weight and, therefore, the cost of the wingsail structure. After that, the design is completely defined at the structural level (including all the auxiliary systems such as the actuators and the control system) and the weight and cost can be used in the Device Assessment Tool.

Another matter of interest is that, once the weights of all the components of the wingsail are known, the centre of gravity of the sail can be computed. This variable is needed for the proper resolution of the stability equations. As explained before, a limitation to the admissible list angle produced by the device is introduced within the VPP and this list angle is closely related to the heeling moment caused by the wingsails. Thus, the position of the centre of gravity of the wingsail will determine the value of the heeling moment produced by the wingsails' weight, which might be relatively high.

However, distinctly from the others, the number of modules is not a previously defined parameter, since it has no effect on the performances of the wingsail. Thus, it can be optimised internally within the Parametric Sizing Tool in order to obtain the minimum-weight combination of modules for the same height, width, etc. Therefore, the tool looks for the number of modules that achieve the minimum weight and cost before passing this data to the Device Assessment Tool.

5. CASE STUDY

For a better understanding of the tools themselves and the importance of selecting the appropriate configuration of the device for each vessel, a case study has been conducted. The study has employed two different vessels: a handysize chemical tanker and a car carrier vessel. For these cases, the economic feasibility of a retrofitting was assessed.

Both have been simulated in a trans-oceanic route between Seattle (US) and Yokohama (Japan) with different device configurations. The wind probability matrix of this route was obtained from 2010 to 2014. This route was selected, among others, by its low wind variability along the route, due to the expository purpose of the study. This feature avoids introducing distorting elements as well as reducing the needed route sections. For the same reason, the route simulation method of the wind probability matrix was used. An example of the obtained wind probability matrix of one section of this route is shown in Figure 4.

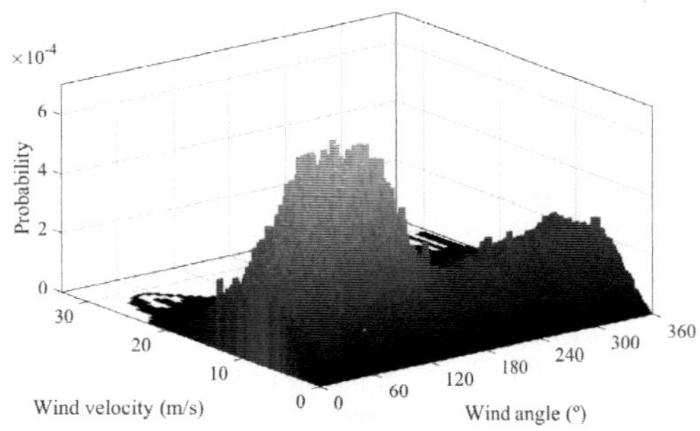

Figure 4: Wind probability matrix of Seattle-Yokohama.

For these generic cases, simplified models of the wingsails' system aerodynamics have been selected. The system is assumed to behave as if it was an isolated infinite aerofoil without taking into account three-dimensional effects or sails interference. Since the configurations simulated have more than one sail, forces are extrapolated multiplying the force of a single wingsail. However, as stated before, the tool admits more complex models (computational or experimental) considering the above-mentioned effects.

The aerofoil selected for the sails was a simple symmetric NACA 0025. Additionally, in all cases a maximum list angle of 5 degrees was established, that will limit the aerodynamic performance of the sails.

Regarding the hydrodynamic forces acting on the hull, the Holtrop method (Holtrop & Mennen, 1978) was used to predict the ship resistance.

With relation to the economic model, some simplification and assumptions were made. In first place, the eventual extra time obtained because of the ship velocity increase (as it is mentioned in *Economic Analysis*) is spent waiting at port for the next scheduled journey, so the economical profit of this extra time is directly earned as a fuel consumption reduction.

In addition, with respect to the operation conditions of the ship, it was assumed that the navigation rate (days at sea per year) was 60% for all cases based on data from (IMO, 2015). The rest of the days are needed for regular maintenance operations and loading/unloading procedures.

Lastly, the OPEX derived from the maintenance of the wingsails' system has been neglected for the sake of simplicity and hence, the total cost (manufacturing and installation) of the wingsails' system for the shipping company has been modelled as:

$$C = (P_{mu} + P_{iu}) \cdot n_s + P_{if}$$

where

> C is the total cost of the system
> P_{mu} is the unit cost of manufacturing
> P_{iu} is the unit cost of installation
> n_s is the number of wingsails of the system
> P_{if} is the fixed installation cost

It must be remarked that, although the absolute value of the systems' prices is a preliminary calculation with a high grade of uncertainty, their relative value is based on the systems' weights and thus is more reliable. This means for the final economic parameters that, although they might change in value, their relative behaviour will remain. At all events, these cases are just general approximations that illustrate the authors' findings and more in-depth analysis should be conducted in real cases.

Thus, the pay-back period and the Net Present Value in a ten-year period have been calculated for each of the sail-vessel combinations as two of the most representative indicators of profitability.

5.1 CHEMICAL TANKER

For the first case, a standard handysize chemical tanker hull was used at a certain load condition. In it, the bound4blue wingsails' system was simulated with two, three and four sails. Each of these cases were run with two possible wingsail's heights, 30 metres and 40 metres. The results obtained for each combination of height and number of sails are shown in Table 1.

About these results three items should be remarked. First, they show clearly how the increase in number or height of the sails does not guarantee any economic improvement (in cases 5 and 6, for instance, fuel savings increase but pay-back and NPV get worse). This is explained by the high unit costs of the 40 metres sails while the savings do not increase so quickly for this ship and route.

Table 1: Results obtained for the chemical tanker

Case	**Height** (m)	**# Sails**	**Mean savings** (%)	**Pay-back period** (years)	**NPV 10 yrs.** (M€)
1	30 m	2	11,36	4,38	0,545
2	30 m	3	16,92	4,14	0,891
3	30 m	4	22,25	4,06	1,207
4	40 m	2	19,80	4,02	1,090
5	40 m	3	27,76	4,14	1,462
6	40 m	4	32,94	4,56	1,456

Second, the lowest pay-back time (case 4) does not match with the highest NPV (case 5). This fact shows how the selection criteria of the best device configuration must be carefully defined according to the financial requirements and long-term profitability prospects.

Third, if cases 2 and 5 are compared, they have almost the same pay-back but, on the contrary, case 5 offers more fuel savings and a higher NPV (about 64% more). Altogether, this suggests that selecting the most appropriate configuration of the wingsails' system is not straight-forward and must be carefully analysed. In other words, there are more economical factors as well as other customer requirements that must be taken into account in each particular case, introducing more accurate models of the system performances and economics.

5.2 CAR CARRIER

For this second case, a general car carrier hull of about 180 m length was used at a certain load condition. For this case, the bound4blue wingsails' system was also simulated with two, three and four sails. However, each of these cases were run with two other wingsail's heights, 20 metres and 30 metres, due to its reduced stability. The results obtained for each combination of height and number of sails are shown in Table 2.

Table 2: Results obtained for the car carrier

Case	**Height** (m)	**# Sails**	**Mean savings** (%)	**Pay-back period** (years)	**NPV 10 yrs.** (%)
7	20 m	2	2,78	3,34	0,585
8	20 m	3	4,15	3,00	0,960
9	20 m	4	5,48	2,85	1,317
10	30 m	2	6,05	2,69	1,514
11	30 m	3	8,26	2,77	2,023
12	30 m	4	9,69	3,05	2,208

In this case the same effects that appear on the chemical tanker can be appreciated. The best payback is obtained for case 10 but the best NPV is obtained for case 12, which is the one with more sails and more height. With this ship, for cases 8 and 12, which have similar paybacks, the difference between the fuel savings offered and the NPV (about 130% more) is even greater.

Additionally, the higher service velocity of car carriers and its correspondent higher hydrodynamic resistance make that the fuel costs considerably increase for this case. This entails that, for lower fuel savings (in percentage), the payback periods and NPV obtained are better even with smaller systems.

6. CONCLUSIONS

Through the contextualisation of the Wind Assisted Propulsion field and the technology transfer problem, including a preliminary analysis of the economic aspects of some different device configurations, it has been illustrated the importance of the correct selection of the system parameters in order to maximise the profitability, which is the key factor for shipping companies.

More specifically, it is shown how larger systems (more sails or more height) could entail smaller benefits and that fuel savings are not a reliable indicator of profitability, as happened with the car carrier case.

7. REFERENCES

1. SHIP & BUNKER TEAM, *World Bunker Prices*, https://shipandbunker.com/prices, 31/07/2017.
2. IMO, *Annex VI of the International Convention for the Prevention of Pollution form ships* (MARPOL), 31/07/2017.
3. IMO, *Prevention of Air Pollution from Ships*, http://www.imo.org/en/OurWork/Environment/PollutionPrevention/AirPollution/Pages/Air-Pollution.aspx, 31/07/2017.
4. IMO, *Third IMO Greenhouse Gas Study 2014*, 2015.
5. IMO, *The Marine Environment Protection Committee*, RESOLUTION MEPC.203(62), 15/07/2011.
6. EUROPEAN COMMISSION, *Transport Emissions*, http://ec.europa.eu/environment/air/transport/ships.htm, 31/07/2017.
7. IMO, *The Marine Environment Protection Committee*, RESOLUTION MEPC.177(58), 10/10/2008.
8. KAZUYUKI O. *et al.*, Third International Symposium on Marine Propulsors, *"Wind Challenger" the Next Generation Hybrid Sailing Vessel*, 05/2017.
9. PALMER, C., GIFFORD and PARTNERS, *Measurements of the Comparative Performance of Sailing Rigs*, 04/1985.
10. NANCE, C., C.ENG, M. & MEDINA YACHT CO LTD. *Outlook for wind assistance.* Journal of Wind Engineering and Industrial Aerodynamics. 07/1985.
11. KERWIN, J., Department of Ocean Engineering, M. I. & Project, H. I. P. O. R. H. A *Velocity Prediction Program for Ocean Racing Yachts* (Massachusetts Institute of Technology, Department of Ocean Engineering), 1975.
12. LARSSON, L. *Scientific Methods in Yacht Design.* Annual Review of Fluid Mechanics 22, 349–385, 01/1990.
13. LEE, H, JO, Y, LEE, D. & CHOI, S., *Surrogate model based design optimization of multiple wing sails considering flow interaction effect.* Ocean Engineering, 2016.
14. HOLTROP, J. and MENNEN, G.G.J., *A statistical power prediction method*, International Shipbuilding Progress, Vol. 25, 10/1978.

8. AUTHORS BIOGRAPHY

David Ferrer Desclaux is co-founder and CTO at bound4blue. He is responsible for the lead and management of the team in charge of the development of bound4blue's technology. He is co-author of three research papers related to the use of the wingsail technology for various applications and co-inventor of three bound4blue patents. David holds a MSc in Aerospace Engineering, with a specialisation in aircraft design and performances (Polytechnic University of Catalonia – Supaero, 2009); and a MSc in Project Management (La Salle – Ramon Llull University, 2016). In 2011 he attended the Space Studies Program (ISU), and has also attended several seminars on start-up management at IE Business School. Previously, he spent 6 years at zero2infinity, right after its foundation, acting as CTO and Project Manager, where he designed, built and tested Near-Space capsules and launched high-altitude balloons from a Spanish Air Force base. He is also co-author in papers related to microgravity's impact on humans and near-space manned flight.

Mario Mantilla Sánchez is performance analyst at bound4blue. He is responsible for the development of the sail-assisted vessel performance analysis tools. He holds a BSc in Aerospace Engineering with a specialisation in aerospace technologies (Polytechnic University of Catalonia, 2017) and is currently pursuing a MSc in Aerospace Engineering (Polytechnic University of Catalonia). As his BSc Thesis, he developed a study for the development of a Weather Routing Software applied to Wind Assisted Propulsion.

Albert García Plaza is structures engineer at bound4blue. He is responsible for the structural and mechanical design of bound4blue's rigid wingsail technology. He holds a BSc in Aerospace Engineering, with a specialisation in aerospace vehicles design (Polytechnic University of Catalonia, 2017) and is currently pursuing a BSc in Physics (UNED), with a specialisation in theoretical physics. As his BSc Thesis, he studied the feasibility of a new kind of airship for merchant transport, which generates lift through vacuum chambers. He worked at Ferrovial Services in his Internship Young Talent program as a trainee in the department of mechanical engineering before joining the bond4blue team.

José Miguel Bermúdez Miquel is co-founder and CEO at bound4blue. He is responsible for the company's management, but providing additional support to the technical development of the technology. Jose Miguel has been active on the topic of using wingsails for multiple purposes since 2006, developing several patents about bound4blue's technology and its use for alternative applications. He is also co-author of three research papers related to the use of the wingsail technology for various applications. José Miguel is a serial entrepreneur. He holds a MSc in Aerospace Engineering, with specialisation in space (Polytechnic University of Catalonia, 2009). He also pursued further education on start-up management at IESE Business School and IE Business School. He joined zero2infinity at its foundation in 2009, the first European company offering private spaceflights using stratospheric balloons. He was involved with the team in business development and engineering tasks. Before joining zero2infinity, Jose Miguel also worked for GMV Aerospace & Defense, in Madrid, designing tools for trajectory simulation of aircraft and NOTAM generation, as well as in GPS signal procedures. He is also co-author in papers related to EGNOS signal processing for NOTAM's generation.

Cristina Aleixendri Muñoz is co-founder and product engineer at bound4blue. She is responsible for the aerodynamics and product certification/approval. She is co-author of three research papers related to the use of the wingsail technology for various applications and co-inventor of three bound4blue patents. She holds an MSc in Aerospace Engineering (Polytechnic University of Catalonia, 2014), developing her MSc Thesis on bound4blue's concept. Before founding bound4blue, she worked as researcher for the Polytechnic University of Catalonia, carrying out thermal and electromagnetic interference analysis using FEM software.

WIND POWERED VESSELS WITH HYBRID AND AUTONOMOUS TECHNOLOGIES

K. Goh, Knud E. Hansen A/S, Australia

SUMMARY

Reducing CO2 for shipping is a high priority but making small efficiency gains in current internal combustion technology and using wind assisted systems is unlikely to make the significant changes to greenhouse gases that are being sought.

In circa year 2000 KNUD E HANSEN had developed an impressive concept design for a fully wind powered vessel – The WindShip. Comprehensive research into the economics and sail technology showed that a wind powered vessel for bulk and liquid cargo was feasible on certain trading routes, but in the year 2000 the additional capital expenditure was not able to be recovered due to very low fuel prices.

Today the economic and environmental situations make the WindShip much more attractive. The new WindShip 2025 concept combines hybrid diesel, solar and battery technologies to enable the WindShip to be even more economical. Autonomous and unmanned operations reduce operating costs and simplify the ship design and systems required. Cargo capacity is also increased.

A new calculation model combines the performance of the sail rig and hybrid power system as well as operational costs and was benchmarked against the previous studies to ensure accuracy. The results demonstrate that large savings in emissions and the cost of operation are now possible with a WindShip that can be developed with today's technologies.

PREVIOUS STUDIES

In 1996-1998, KNUD E HANSEN participated in a comprehensive wind powered ship study sponsored by the Danish Environmental Agency. The study involved both technical and economic feasibility. The study was performed in two phases and settled on a 50,000 DWT bulk carrier initially and a product tanker in the later phase.

An exhaustive review of sail rig technology was undertaken and a novel and powerful high-lift wing sail was developed that offered impressive performance. The wing had a very high lift coefficient of about 3.0, compared to the DynaRig of about 1.5. This meant that the required sail area could be vastly reduced.

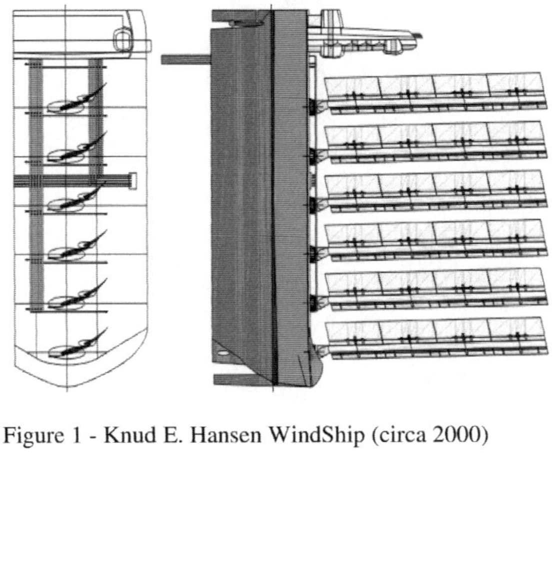

Figure 1 - Knud E. Hansen WindShip (circa 2000)

The sail rig consisted of a number of wing sections on each mast. In order to tack the wing sections would rotate about the mast. The wing could also be reefed in case of extreme wind conditions or during bulk loading operations.

The high-lift wing was able to sail much closer to the wind than other square type rigs. About 40 degrees was achievable. Another advantage was the robustness of the rigid wing sections. Compared to cloth type sails, that would need replacing every few years due to UV degradation, the wing could be built from composite panels and have the same life time as the vessel.

Figure 2 - High lift wing sail changing direction

Technical & Economic Analysis

The technical study was extensive and analyzed:

- Sail rig performance
- Wing & mast design
- Hull form & resistances
- Hydrostatic stability
- Longitudinal stability under sail

In a favorable wind, the vessel could sail at speeds close to 20 knots. The vessel was also provided with twin azimuthing thrusters and a bow thruster for auxiliary propulsion and docking without tug assistance.

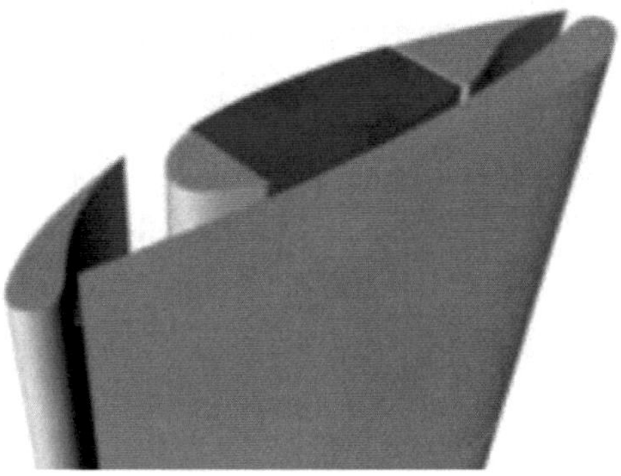

Figure 3 - Wing section reefed

An economic study was undertaken by Maersk Brokers which examined two product tanker trading routes and compared the WindShip to a conventional tanker with 2-stroke motor propulsion. The WindShip was only economically viable on some of the trading routes chosen, while using up to 10% more fuel on others equatorial routes where the average wind speed was only about 5m/s.

Speed/ Consumption	11[knots]	12 [knots]	13 [knots]
Conventional Ship [tonnes/24hr]	17	23	30
Windship [tonnes/24hr]	14.2	20.0	27.1
Difference	16.5%	13%	9%

Table 1 - Convention vs WindShip on Atlantic trade

Athough the fuel consumption was reduced, the poorer economic performance of the WindShip compared to conventional motor ships was attributed mainly to the speed being fixed at a relatively high speed of 13 knots to provide a direct comparison with existing vessels and poor wind conditions on the largely equatorial trading routes chosen. Other factors included the lower propulsive efficiency of azimuth thruster system and the higher capital costs of the WindShip.

REVISITING THE WINDSHIP

With a renewed focus on reducing CO_2 emissions and the coming global cap of 0.5% sulphur bunker fuel, the economic and political climate is now looking much more positive for the return of wind power ships.

In addition, maturing technologies in the fields of solar power, battery storage, fuel cells and autonomous control, provide further reasons to move on from the traditional motor ship.

The purpose of this study is to take the promising WindShip concept developed in the late 90's and see if the concept can be justified economically for the year 2025, the proposed in-service date for the new WindShip, the WindShip 2025.

New Calculation Model

The first step was to develop a new calculation model that would be able to quickly verify the basic economics of the WindShip and compare them to a conventional motorship. The model should be able to calculate sailing performance by wind and auxiliary propulsion systems for any chosen sailing route and thereby also inform the fuel consumption based on generally accepted efficiencies.

The 50,000 DWT Handymax bulk carrier studied previously would serve as the basis of the new WindShip calculations, since the performance was already known from extensive model testing. Although further performance optimizations to the sail rig were possible from the knowledge gained from the previous study, it was decided to utilize the data set as closely as possible to ensure accurate benchmarking of the new calculation model.

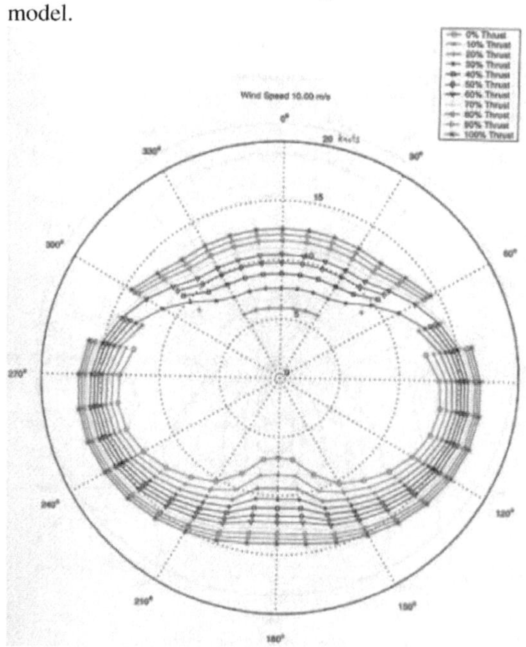

Figure 4 - WindShip performance at 10m/s w

50,000DWT Bulk Carrier	**Conventional**	**WindShip (year 2000)**
Length BPP	210 m	210 m
Beam	32 m	32 m
Draft	11.9 m	11.9 m
Block	0.77	0.75
Lightweight	11,500 t	12,000 t
Cargo Capacity	69,000 m^3	71,000 m^3
Propulsion efficiency	0.6	0.6
Installed prop'n power	8,000 kW	6,000 kW
Prop'n power SFC	0.18 kg/kWh	0.2 kg/kWh
Design speed	14.0 kn	13.0 kn
Typical elec. load	465 kWe	400-500 kWe
Aux. power SFC	0.2 kg/kWh	0.2 kg/kWh

Table 2 - Conventional vs WindShip main particulars

The speed and resistance from model testing of the hull and sail rig was translated from the polar plots of wind speed and direction into a spreadsheet. From the resistance of the hull and sail rig, the effective power and theoretical shaft power was calculated from the estimated propulsion efficiency of 60%.

Two common dry bulk routes were chosen for analysis. The Trans-Atlantic route from Rotterdam to New York City and ballast return was selected since it was known that this route has good average wind conditions of 8m/s. The other route chosen was a long Trans-Pacific grain trade route from Vancouver to Shanghai with ballast return. It is known that the wind conditions on this route results in a lower wind average.

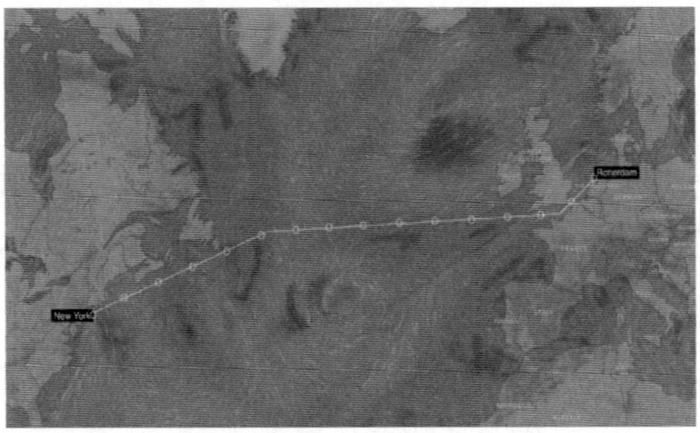

Figure 5 - Wind data on Atlantic trade route

Global wind data was sourced from the European Centre for Medium-range Weather Forecast (ECMWF). The vessel course was plotted directly and divided into 16 legs. Three trials for each route were calculated and averaged for three different speeds, 8, 11 and 13 knots. Weather routing was not applied.

From the vessel course, the relative direction of the wind was calculated for each leg and from the wind speed, the vessel speed was known. If the vessel was pointing higher than 45 degrees into the wind, the vessel would tack and the additional distance from the tacked course was calculated for that leg. Additional electrical load was also allowed for turning the wing sections when changing tack.

Benchmarking Calculations

For the benchmarking calculations, the WindShip was generally allowed to run as fast as the wind would allow. Auxiliary propulsion was used if the sail power was insufficient to make the required average speed. Sea margin was also added based on the wind speed.

The WindShip fuel consumption was calculated based on the auxiliary propulsion requirement and the service electrical load. The return voyage at ballast condition was calculated at 85% of the deadweight fuel consumption. The price of fuel used for the benchmarking calculation was 200 USD/ton for HFO and 300 USD/ton for MDO.

A simple regression based cost model was used to calculate new building costs based on steel and equipment weights, block coefficient, installed power and machinery type. The building cost of the total sail rig was estimated at 5.0 MUSD for year 2000, which resulted in a 23% increase over the conventional motor ship cost. The capital financing period was 10 years. Operating costs were taken from the previous study, which was based on year 2000 costs for benchmarking. Manning, maintenance, stores and administration costs where the same as the motor ship. Insurance costs reflected the increased cost of the WindShip and port dues were reduced since tugs would not be needed for docking.

The results calculated with the new economic model were very close to the previous study. Like the 1998 study, the model showed that a WindShip was not economically viable in the year 2000, in principle because at 13 knots speed, the marginal saving on fuel

ROT-NYC T3	3300																			
Leg	Distance	Course	Wind dir.	True Wind	Wind Speed	Sea Margin	Tack	True dist.	Speed by wind	Speed req'd	Drive mode	Prop'n Power	Aux. Power	Recov'd Power	Solar Power	Net Power	Net Energy		Duration	
	deg	deg	deg	deg	m/s			nm		kn		kWe	kWe	kWe	kWe	kWe	kWh		hr	
1	206	230	0	230	20.0	0.60	NO	206	19.0	18	HARVEST	0	90	941	172.8	-1023	-11,726		11.5	
2	206	230	345	115	12.5	0.38	NO	206	17.0	16	HARVEST	0	90	622	172.8	-705	-9,092		12.9	
3	206	265	0	265	7.5	0.23	NO	206	10.5	12	ASSIST	1948	90	0	172.8	1865	32,061		17.2	
4	206	265	45	220	5.0	0.15	NO	206	5.0	8	ASSIST	967	90	0	172.8	884	22,796		25.8	
5	206	265	135	130	7.5	0.23	NO	206	10.0	8	HARVEST	0	90	400	172.8	-483	-12,443		25.8	
6	206	265	145	120	10.0	0.30	NO	206	13.0	12	HARVEST	0	90	365	172.8	-447	-7,691		17.2	
7	206	265	135	130	7.5	0.23	NO	206	10.0	10	HARVEST	0	90	0	172.8	-83	-1,708		20.6	
8	206	265	350	85	20.0	0.60	NO	206	18.0	17	HARVEST	0	90	1659	172.8	-1741	-21,128		12.1	
9	206	265	315	50	12.5	0.38	NO	206	8.5	9	ASSIST	553	90	0	172.8	470	10,774		22.9	
10	206	265	315	50	7.5	0.23	NO	206	6.0	7	ASSIST	291	90	0	172.8	208	6,137		29.5	
11	206	245	150	95	5.0	0.15	NO	206	7.0	7	HARVEST	0	90	0	172.8	-83	-2,440		29.5	
12	206	245	160	85	15.0	0.45	NO	206	16.0	15	HARVEST	0	90	597	172.8	-680	-9,345		13.8	
13	206	245	190	55	7.5	0.23	NO	206	6.0	6	HARVEST	0	90	0	172.8	-83	-2,846		34.4	
14	206	245	220	25	7.5	0.23	YES	242	6.0	6	HARVEST	0	120	0	172.8	-53	-2,133		40.4	
15	206	245	250	5	7.5	0.23	YES	242	6.0	6	HARVEST	0	120	0	172.8	-53	-2,133		40.4	
16	206	245	270	25	15.0	0.45	YES	242	9.0	8	HARVEST	0	120	281	172.8	-334	-10,114		30.3	
	3,300				10.5	0.31		3408		8.9						Net Energy	-21,029	kWh	384.1	hrs
											Conv Ship	Pd (kW)	Paux (kW)			Fuel WS	-5.0	t	16.0	days
												2011	565			Fuel CS	220.3	t	102%	

Table 3 - Calculation model on Atlantic trade at 11 kns

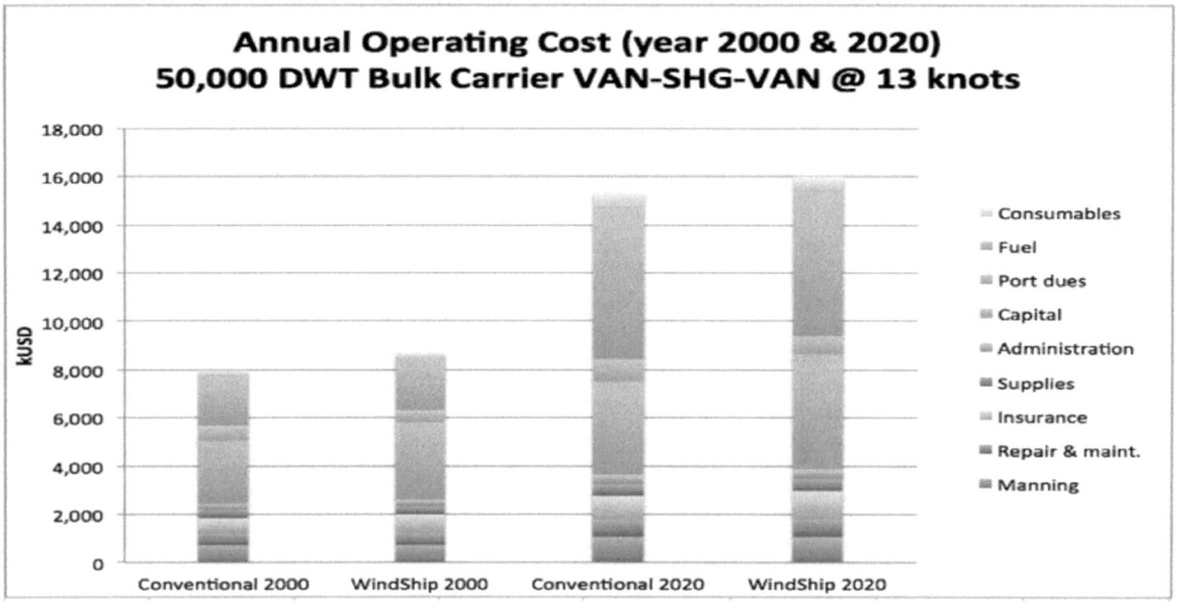

Figure 6 - Conventional & WindShip annual costs for year 2000 & 2020

consumption that the sail rig provided was out weighed by the less efficient auxiliary propulsion system, the higher consumption for medium-speed engines and the higher capital cost of the WindShip.

The results showed similar or better operating costs for the conventional bulk caririer and the WindShip on the Rotterdam – New York City route where the average wind speed was 8.6 m/s and the fuel saving could offset the higher capital costs. However, the Vancouver-Shanghai route where the average wind was only 6.7 m/s had a 10% higher operating cost.

The cost model was then updated to represent costs for year 2020 to check the economic viability of WindShip built today. Annual inflation of 2% was applied to both capital and operating costs. The fuel price increased to 600 USD/ton for 0.5% low sulphur HFO and 800 USD/ton for 0.1% sulphur MDO. The calculations showed that the WindShip was now only 4% more expensive to operate on the Vancouver-Shanghai route at 13 knots and would likely have the economic advantage at lower speeds.

THE WINDSHIP 2025 CONCEPT

With the tipping point for economic viability likely to be triggered by the IMO global 0.5% bunker fuel sulphur cap in 2020, it is now the time to see what the WindShip 2025 concept could achieve.

Auxiliary Propulsion

Although sailing ship once plied the world's oceans powered only by the wind, this is not desirable for the

WindShip. An effective auxiliary propulsion system for the Windship will allow the vessel to:

- Maneuver in and out of port unassisted
- Assist in course keeping and improving the efficiency of the sail rig
- Sail when no or little wind (<2.5m/s) is available
- Assist the sail power to keep to a schedule
- Take-home-power should the sail rig become compromised

Figure 7 - WindShip 2025 Concept

Another benefit of the new auxiliary propulsion system is the possibility of using the system as a wake-turbine generator when there was excessive wind. A number of different energy storage systems were considered including flywheels and electrolysis of water into hydrogen, however the simplest and most mature technology with a high round-trip efficiency is lithium-ion batteries. This would necessitate using an electric propulsion drive system.

The new WindShip has an auxiliary propulsion system using two aft azimuthing electric pods with controllable pitch propellers. The calculation model was modified to allow for energy storage from a Power-Take-In and Power-Take-Out propulsion system utilising lithium-ion batteries. The turbine efficiency was estimated at 50% of the difference of effective powering requirements for the vessel speed by sail and the harvesting speed for the known wind speed. The stored power could then be used when wind conditions were not favourable or when a higher speed was needed to meet the sailing schedule. In favourable wind and a harvesting speed of 16 knots, about 2,000 kWe can be generated and stored. A power of 2 MWe is enough to power the WindShip at 11 knots and run all auxiliaries when there was no wind to sail.

Unmanned & Autonomous Operation

The WindShip 2025 will be an unmanned vessel, the sail rig being ideally suited for mechanisation and automated control. This also has the possibility to eliminate all life-saving equipment, hotel systems and large associated electrical loads such as HVAC and lighting. Removing the vessel superstructure also reduces the aerodynamic drag and improves the performance of the sail rig.

To reduce the complexity further and to reduce breakdown and fire risks for an unmanned vessel, the WindShip 2025 also has no engine room. Generator machinery and fuel storage is containerised and can be carried according to the requirements of the route. High-speed diesel generators are not as efficient and distillate MGO fuel is more expensive but this was necessary to work with the PTO-PTI propulsion system and the unmanned concept.

Figure 8 - Containerised Power System

A standard forty-foot container module is able to house a 3,000 kWe diesel generator, so only two would be needed for the 13 knot design speed. A forty-foot ISO tank is able to store about 40 tons of fuel. Ten to twelve tanks are enough for almost any voyage up to 5,000 nm at 13 knots.

Energy Storage System

The lithium-ion battery is one of the key technologies that is enabling a clean and energy efficient future. With excellent cycle capability and a high round-trip-efficiency of about 90% they are an obvious choice for the WindShip 2025 concept.

The previous high cost of lithium-ion batteries has been dropping quickly, driven by the increasing maturity and uptake of electric vehicles and on-site energy storage. From a price of over 1,000 USD per kWh in 2010, *Bloomberg* report that the cost of Li-ion batteries is reducing at 21.6% per year. If this trend continues, a cost of under 100 USD per kWh is certainly possible by 2025. To allow for the containerised housing system, a conservative cost estimate of 150 USD/kWh is used.

The batteries system would also be containerised. Each forty-foot ISO container can conservatively store 2,500 kWh of lithium-ion batteries with adequate room for cooling and power management systems. The storage limit would likely be determined by the container weight limit of 30 tons.

Battery Config'n	**Number of 40ft ISO units**	**Storage Capacity**	**Cost**	**Range at 8 kn**	**Range at 11 kn**	**Range at 13 kn**
Base	10	25 MWh	3.75 MUSD	133 nm	78 nm	54 nm
Maximum	40	100 MWh	10.0 MUSD	532 nm	322 nm	216 nm

Table 4 - Battery Configuration

How much storage should be provided is the other battery factor. A 2.5 MWh battery container can only power the WindShip at 6 knots for 3 hours, 8 knots for 1.5 hours and 10 knots for one hour. Powering is only half of the storage calculation with harvesting being just as important. In a good wind of over 15m/s the sail rig could propel the WindShip at 19 knots, which is an effective power of 11 MW. If the harvesting speed is 17 knots (effective power of 7.5 MW), then the harvested power is 1.7 MWe and able to fully recharge a 2.5 MWh battery container in about 1.5 hours.

It was decided to trial two different battery configurations. A base battery configuration of 10 containers with a 25 MWh storage capacity and a maximum battery configuration of 40 containers with 100 MWh.

Solar Power System

Solar power through photovoltaic cells is another key technology that is enabling a clean and energy efficient future. Lightweight and modular, they are also an obvious choice for the WindShip 2025 concept. Although the energy density is still very low at about 200 Watts per square meter, it is envisaged that solar panels can be fitted to the upper surface of the wing sections and virtually all flat deck areas. It is estimated that the vessel can be installed with 7,200 square meters of solar panels producing a peak power of 1,440 kWe, enough power to sail at 7 knots and run all service loads at the same time.

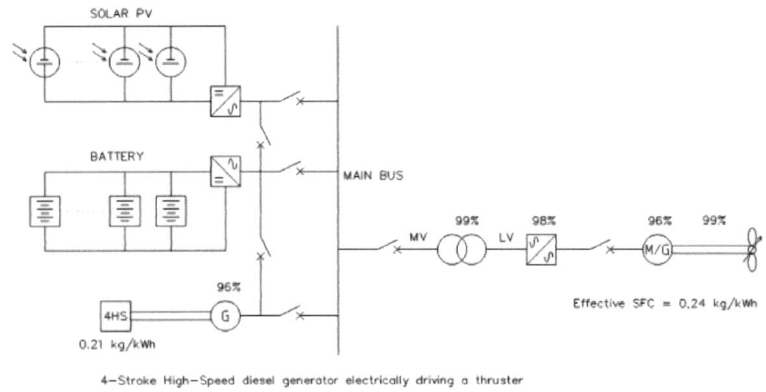

Figure 9 - WindShip 2025 power system

When there are periods of very little wind, or when the WindShip 2025 is at anchor or in port, the sails can track the sun to maximize energy harvesting and recharging of batteries. Even if the vessel is sailing by the wind, solar radiation from reflection off the sea surface can still provide a significant amount of energy. The calculation model uses a tracking factor of 0.7 and an untracked factor of 0.3 and also allows for a 40% daylight factor to allow for cloud and nighttime.

The previous high cost of photovoltaic cells has been dropping quickly, driven by the increasing maturity and global uptake of solar power both on-site and mass energy harvesting and distribution. From a price of about 1100 USD per kW in 2010, Bloomberg report that the cost of photovoltaic cells is reducing at 24.3% per year. If this trend continues, a cost of under 50 USD per kW is possible by 2025. Since a large part of the cost will be installation and support structures for the solar panels a much more conservative cost of 500 USD per kW has been used.

Fuel Cells

A fuel cell device combines hydrogen and oxygen to make electricity with water and heat being the waste products. The main benefits are the simplicity with no moving parts, higher energy efficiencies compared to internal combustion engines and less-polluting waste products. Although use of fuel cells for ship propulsion is in its infancy, the concept works well with the unmanned WindShip and large scale implementation for shipping may be ready by year 2025. DNV recently undertook a detailed fuel cell study looking into the technology, safety aspects and regulatory requirements in preparation for wide spread use fuel cells in the near future.

The efficiency of the fuel cell itself is currently between 45-60% depending on type, the rest of the energy being lost as waste heat. With high temperature fuel cells, the waste heat has the potential to be recovered through a steam turbine or a Stirling Engine, increasing the overall efficiency to 85%. An additional benefit of the high temperature fuel cells are the ability to directly use LNG and converting to hydrogen through a process called internal reforming.

For the WindShip 2025, the fuel cells would be containerised and replace the generator engine modules. Containerised LNG tanks would be carried instead of the diesel tanks. The fuel cell version of the WindShip 2025 is also included in the economic study. A factor of 1.5

was applied to the machinery costs to reflect the likely higher costs of the fuel cells, which will not have reached high levels of economy of scale.

WindShip 2025 Costs

The cost model allowed for the reduction of about 500 tons of steel by removing the superstructure and engine casing. Equipment costs was also reduced by 500 tons since there was no hotel services, no lifesaving equipment or engine room. A pump room for ballasting control is still required and switchboard rooms for distribution of power and housing of the propulsion drives. Machinery costs were calculated based on 4-stroke power, plus a 1.25 factor for a diesel-electric system, 1.5 factor for azimuthing pods and 0.6 factor for the containerised generators set. An unmanned automation cost of 1 MUSD was also added to the capital cost of the WindShip.

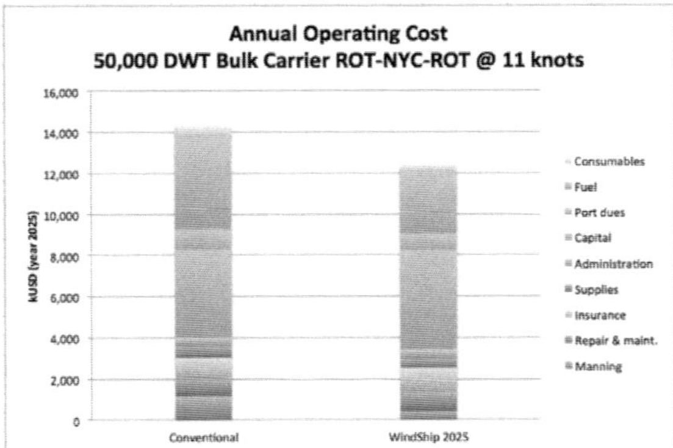

Figure 10 - Annual cost for Atlantic trade at 11 kns

Regarding annual operating costs, manning was reduced to a third of the motorship, since the unmanned WindShip would still require crew to command and monitor the vessels from a shore based control station. It is foreseeable, that is the future there will be only a few people in charge of an entire fleet of ships operating autonomously 99% of the time, however traditional jobs will still remain, including inspection and maintenance of the WindShip when at anchor or docked.

Maintenance cost was increased to reflect the needs of the sail rig and the fact that there was no crew aboard to cater for daily maintenance tasks. Insurance costs of 2.5% also reflected the increased capital costs of the vessel. Administration costs where also increased slightly the increased to reflect the real-time video and control communication system needed for autonomous operations. Annual inflation of 2% was applied to both capital and operating costs. The fuel price increased to 800 USD/ton for 0.5% low sulphur HFO and 1000 USD/ton for 0.1% sulphur MDO.

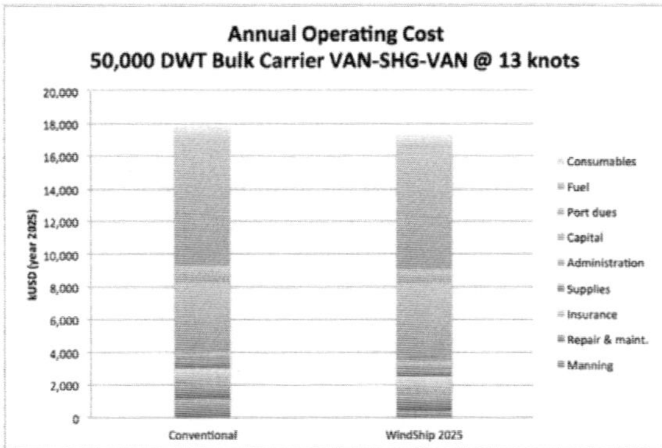

Figure 11 - Annual cost for Pacific trade at 13 kns

CALCULATION RESULTS

The calculations showed that the WindShip 2025 concept was economically viable on both the Rotterdam - New York City route and the Vancouver – Shanghai route at all speeds of 8, 11 and 13 knots. A total operational cost saving of about 15-20% can be expected with the WindShip 2025 compared to a conventional motor ship.

The Required-Freight-Rate was best for both the conventional motor ship and WindShip 2025 at about 11 knots. There was a saving of only a 2-3 percent at 13 knots on the Vancouver – Shanghai route, but the CO2 saving was still considerable.

The shipbuilding CO2 was also added to the operational CO2 to a get the complete environmental impact and calculated at 5 tons of CO2 for every ton of the ship lightweight and the total pro-rated over a 25 year vessel life.

It can also be seen that the 'max-battery' WindShip was not able to recover the higher capital cost on either route at any speed.

The fuel cell Windship with its much higher efficiency compared to the diesel generators, was by a significant margin, showing the best economic and environmental performance. However, due to immaturity of the technology and as-yet largely unknown costs of large scale implementation, the results presented in this report should be interpreted with caution.

When the results are presented in a hypothetical trade scenario of shipping 5 million tons of grain per year for a conventional motor ship and the WindShip 2025, the implications for fleet size and operational costs become more apparent.

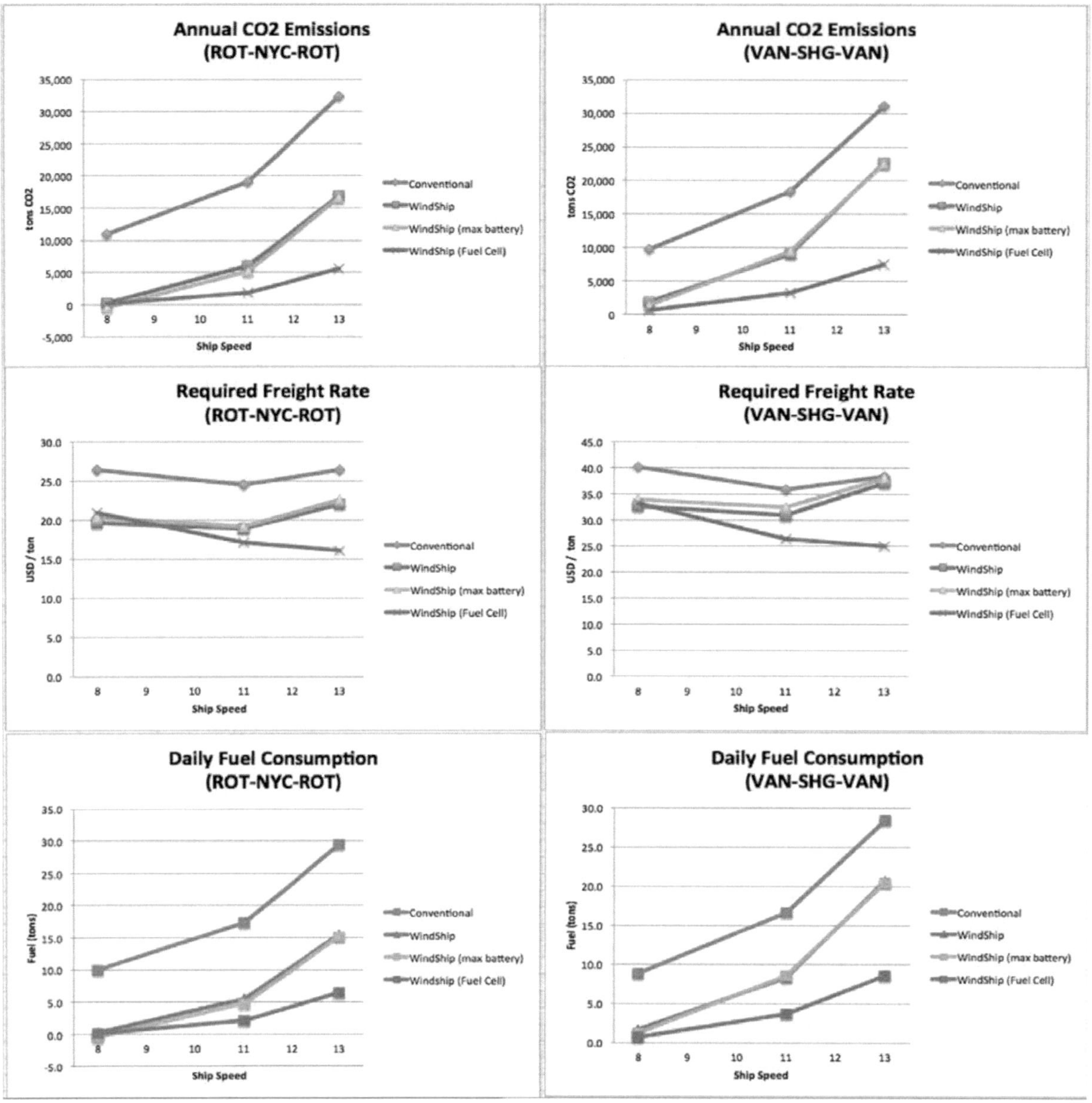

Figure 12 - Performance on Atlantic & Pacific trade routes

CONCLUSIONS

Lower speeds than 13 knots and larger fleets must be accepted to reduce operating costs and environmental impact.

An average wind speed of at least 6.5 m/s is required if a 13 knot service is to be economically viable. The WindShip will not be economically viable on equatorial routes.

Despite using less efficient high-speed diesel generators and more expensive distillate fuel, the WindShip is still more cost effective and less polluting than a conventional 2-stroke motor ship.

About 15% higher capital costs for shipbuilding must be accepted for the WindShip 2025. This does not include R&D and prototyping costs.

The cost of battery storage needs to drop below 50 USD/kWh for energy harvesting to be cost effective.

ROT-NYC-ROT	Convent'l @ 8 kn	Convent'l @ 11 kn	Convent'l @ 13 kn	WindShip @ 8 kn	WindShip @ 11 kn	WindShip @ 13 kn
Grain per annum (tons)	5,000,000	5,000,000	5,000,000	5,000,000	5,000,000	5,000,000
Grain/year/ship (tons)	466,291	593,737	689,034	469,595	598,347	694,669
Ships required	11	9	8	11	9	8
Operating cost per year (kUSD)	12,305	14,534	18,181	9,180	11,280	15,276
Total annual cost (MUSD)	**135**	**131**	**145**	**101**	**102**	**122**
Cost saving	**0**	**0**	**0**	**25%**	**22%**	**16%**
Shipbuilding CO2 (tons)	25,300	20,700	18,400	25,300	20,700	18,400
Operation CO2 (tons)	120,246	171,346	258,385	3,185	54,152	135,308
Total annual CO2 (tons)	145,546	192,046	276,785	28,485	74,852	153,708
CO2 saving	**0**	**0**	**0**	**80%**	**61%**	**44%**

VAN-SHG-VAN	Convent'l @ 8 kn	Convent'l @ 11 kn	Convent'l @ 13 kn	WindShip @ 8 kn	WindShip @ 11 kn	WindShip @ 13 kn
Grain per annum (tons)	5,000,000	5,000,000	5,000,000	5,000,000	5,000,000	5,000,000
Grain/year/ship (tons)	298,212	400,484	466,231	298,622	401,256	467,294
Ships required	17	13	11	17	13	11
Operating cost per year (kUSD)	11,960	14,319	17,850	9,752	12,407	17,363
Total annaul cost (MUSD)	**203**	**186**	**196**	**166**	**161**	**191**
Cost saving	**0**	**0**	**0**	**18%**	**13%**	**3%**
Shipbuilding CO2 (tons)	39,100	29,900	25,300	39,100	29,900	25,300
Operation CO2 (tons)	164,511	237,324	342,061	31,449	118,160	248,662
Total annual CO2 (tons)	203,611	267,224	367,361	70,549	148,060	273,962
CO2 saving	**0**	**0**	**0**	**65%**	**45%**	**25%**

Table 5 - Fleet economics for Atlantic & Pacific bulk trade

FUTURE STUDIES

The high-lift wing sail rig is to be optimised further for performance and cost. For example a three section wing instead of four will be investigated.

Future study to also include other sail systems such as Flettner rotors which have a very high lift coefficient and a simpler construction compared to the WindShip high-lift wing sail. The disadvantage of Flettner rotors is the reduced effectiveness in head and following winds.

A more detailed performance model will be developed to allow weather routing and better modeling of the PTO-PTI energy system.

Fuel cells provide direct conversion of fuel into electricity and high total efficiency has been demonstrated. As the technology matures more accurate modeling is required to ensure that the economic and environmental benefits can be realised.

The azimuthing propulsion pods required for WindShip 2025 do not currently exist. The pods require large controllable pitch propellers that can cover the full range of propulsion and harvesting efficiently at different vessel speeds. The propulsion motors will need multiple windings to ensure the most efficient supply and harvesting of energy at vastly different power levels.

Alternative arrangements for the azimuthing pods will be investigated. For example a Counter-Rotating-Propeller arrangement with just a single azimuthing pod could have advantages.

The economic arguments here do not consider full environmental and social impact of fossil fuels. *Siemens* has investigated such factors in a recent study to support the wind industry. Such impacts and associated cost should be included in any future economic study.

REFERENCES

Modern Wind Ships – Phase 1, *Knud E Hansen A/S 1997*
Modern Wind Ships – Phase 2, *Knud E Hansen A/S 1999*
EMSA Study On The Use Of Fuel Cells In Shipping – *DNV/GL 2017*

ROLE OF RIM DRIVEN PROPULSORS IN FUTURE ELECTRIC SHIPS

S D A Fletcher, **R Hayes**, **C Morgan**, **A Allan** and **C Carter** Frazer-Nash Consultancy, UK

SUMMARY

Rim-driven propulsors (RDPs) have the potential to play an important role in the propulsion of future electric ships. RDPs are compact, electrically driven propulsion systems which can provide flexibility in propulsor location and provide an opportunity to develop alternative ship propulsion architectures. This paper discusses key challenges associated with the use of RDP systems in future ships, with a focus on the potential for RDPs to fulfil the role of main propulsion systems. As part of this analysis, the paper identifies the expected propulsion power levels of which RDPs need to be capable and the potential barriers for meeting these levels.

1. INTRODUCTION

The shipping industry has experienced a significant technology shift in recent years with increasing use of electrical propulsion systems. Electrical propulsion, supported by a substantial change in the ship electrical system, has allowed individual ship operators to experience benefits such as reduced maintenance and reduced fuel consumption.

At a whole-industry level, greenhouse gas emissions from shipping is an area receiving increasing scrutiny and governmental pressure. Maritime transport is currently responsible for about 2.5% of global greenhouse gas emissions, a figure which is expected to increase over time. The European Commission [1] recognise that advanced system design and technology implementation can play a key role in managing these emissions, with reductions in the order of 75% achievable. This presents an opportunity for the further exploitation of low carbon electrical propulsion designs and technologies, such as the rim-driven propulsors (RDPs).

RDPs are electrical propulsion systems that integrate an electric motor within a ducted propeller. A number of potential benefits are offered by RDPs for a range of applications because they remove the need for conventional mechanical drivetrains and offer alternative platform arrangements. Whilst currently employed as COTS devices in smaller scale applications, a number of challenges exist in the scaling of these technologies. These must be overcome to enable benefits in large scale propulsion systems.

The following sections of this paper will first discuss why electrical propulsion can be advantageous in general and the power levels in current operation. This helps to illustrate what RDP systems must be capable of if they are to fulfil the role of propulsion systems on future ships. Later sections go on to describe key challenges associated with integrating RDPs. Areas of interest include the impact of such systems on ship architecture and electrical integration challenges.

2. ELECTRICAL PROPULSION

The use of electrical propulsion is becoming more prevalent on a range of ship types and sizes. Examples of these include:

- Naval vessels (Type 45 [2], Type 26 [3], Queen Elizabeth Carrier [4], Zumwalt [5]);
- LNG supply vessels [6, 7];
- Cruise ships [8, 9];
- Offshore supply vessels [10] and
- Ferries [11, 12].

There are three general configurations of electrical propulsion which ships may employ. These are parallel hybrid, series hybrid or fully electric propulsion systems. Within a parallel hybrid configuration, both an electric propulsor and an engine can provide propulsive power by connecting to the same shaft. This hybrid propulsion arrangement can provide operational efficiency benefits while retaining elements of traditional propulsion system operation [13]. Given that this configuration necessitates the use of a mechanical shaft, it is not applicable to an RDP system.

Fully electric systems can use energy storage (or potentially more novel electrical power generation technologies) to provide power to the propulsor. These are currently limited to short duration, lower power applications, such as ferries [12], given the limitations of available electrical power storage (e.g. battery storage).

A series hybrid configuration is utilised in the majority of platforms described as having electrical propulsion. This configuration generates electrical power from an engine, usually a gas turbine and/or diesel engine for ship applications. Power is then distributed to the electrical motors that make up the propulsion system. A common electrical supply is often used to also supply ship services. This combined supply arrangement can lead to complex and highly integrated electrical systems. This can bring its own challenges for control and system management (e.g. the well-published issues experienced on the Type 45 [2]) as well as many opportunities for optimisation. These include rating of equipment, reliability and operational flexibility.

2.1 ADVANTAGES OF ELECTRICAL PROPULSION

In the series hybrid system configuration electrical propulsion has a number of potential benefits. These have supported the increased uptake in its application over recent years.

In terms of power generation, by breaking the mechanical connection between the prime mover (e.g. a diesel engine) and the propulsor, it enables prime movers, and associated electrical generators, to be located around the ship. Noise and vibration can also be reduced by removing prime movers from the base of the ship. The location of propulsors is also made more flexible by decoupling them from the prime mover.

Electrical propulsion provides the ability to use smaller distributed prime movers, enabling more optimal use across the operating profile by matching operating prime movers to demand. This configuration can also support greater system redundancy and availability.

These benefits combine to offer a more flexible ship design and propulsion system that reduces emissions, fuel consumption, mass, volume, noise and vibration levels and maintenance costs [13].

An extended benefit of the flexibility of location offered by electrical propulsion is that it may facilitate the easier integration of contra-rotating propellers (CRP). A CRP design uses two propellers, placing one in front of the other with each rotating in the opposite direction. The main purpose of this arrangement is to allow the recovery of rotational losses from the forward propeller. Based on the analysis of an example application, (Hoorn, *et al*) estimates that the increase in propulsive efficiency offered by CRP can reduce fuel consumption by around 8%. The flexibility offered by RDP systems may enable greater deployment of such system concepts.

2.2 PROPULSION POWER REQUIREMENTS

The propulsive power required on a given platform will have a significant impact on the suitability and availability of rim-drive technology. A number of factors influence propulsive power requirements, including ship size and top speed requirements. Speed is related to power through a cubic relationship (Lu *et al*, 2015) and therefore small increases in speed can greatly increase propulsive power requirements.

Some examples of how propulsive power varies across different applications are:

- Offshore supply vessel propulsive power is of the order of 4 MW, with main propulsors delivering around 2 MW each (there is a larger variability given the number of platform configurations) [10];
- Cruise ships typically use propulsors rated between 7.5 MW and 15 MW [8, 9];
- The Yamal LNG tanker uses three 15 MW propulsors [7];
- The Type 45 ship and the QEC both use two 20 MW propulsion units (two on the Type 45 and four on the QEC) [2, 4];
- The Zumwalt ship is designed with 70 MW of propulsive power, provided by two 34.6 MW motors [5].

These existing applications help to assess the required power outputs of RDPs if they are to replace current electrical propulsion systems.

3. RIM-DRIVEN PROPULSION

While electrical propulsion is now well established, opportunities exist to further improve these platforms in areas such as controllability, efficiency and reliability. RDPs are one such emerging technology under investigation for this purpose.

3.1 RDP DESIGN AND ADVANTAGES

An RDP is an electrically driven propeller where the stator of the electrical motor is housed within the thruster duct, and the machine rotor forms a ring around the tips of the propeller. The propulsor blades are rigidly attached to this ring. Many design variants are available including both hubless and hubbed designs, illustrated in Figures 1 and 2 respectively (Van Blarcom, Hanhinen, & Mewis).

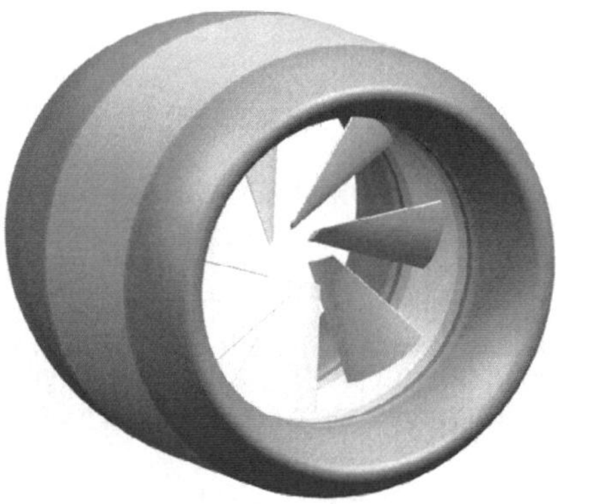

Figure 1: Hubless Rim Drive Design

A rim-driven motor can technically be driven by any electrical machine type (e.g. DC or induction motors), but there is a general preference for permanent magnet synchronous machine types, as reflected in the commercial offerings available. In this configuration, a ring of permanent magnets are mounted on a rotating rim as Figure 3 illustrates (Carter, 2017).

There are a number of potential advantages in the use of RDPs over more conventional electrical propulsion. At the individual propulsor level, RDP systems can reduce acoustic noise signatures due to a reduced propeller tip speed (note however that there are a range of sources of noise and calculation of this can be complex (Carter, 2017). RDP motors typically have large diameters compared to conventional machines, allowing a higher number of pole pairs to be used in the motor.

Figure 2: Hubbed Rim Drive Design

Weight can be removed from the propeller due to the removal of the shaft. This can reduce inertia enabling more rapid control of the rotational speed. The significance of this will depend on the size of the propulsor.

Certain designs can be cooled by the surrounding water and do not require a separate cooling system for the RDP (Van Blarcom, Hanhinen, & Mewis; Hamson & Ward). However there are limits to the cooling this can provide and additional cooling may still be required for high power RDP units due to high stator currents and the associated resistive losses.

The full benefits of RDP technology are seen when considered at a whole ship level. The removal of the shaft allows for units to be positioned with more freedom, which opens up the opportunity for novel configurations to provide enhanced platform capability. Examples include multiple unit primary propulsion, secondary propulsion, dynamic positioning or hover control.

The potential impact of RDPs on hydrodynamic efficiency is one key area of interest. Initial published research based on a scaled cruise ship model estimated that an efficiency improvement in the order of 7% was achievable (Hamson & Ward). This level of benefit is also reflected in the information published by Rolls-Royce on its rim-driven thruster technology [19]. This is significant as it not only reduces the total energy consumed from ship propulsion, but also reduces the peak power requirements. This can enable the total installed generation power on the ship to be reduced along with the associated infrastructure, saving cost and space. Such an improvement would facilitate a reduction in ship size or an increase in payload. Alternatively, higher speed operation could be achieved for the same installed power.

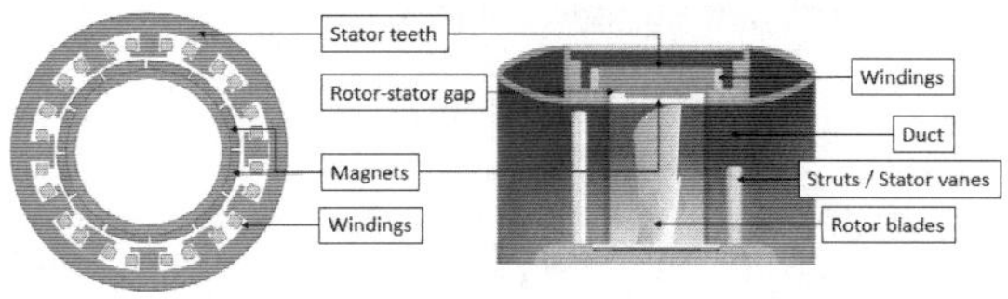

Figure 3: Meridional and Axial Cross Sections through a RDP

Hydrodynamic effects arise due to the interaction of the hull and the development of the boundary layer. The flow into and around the propeller is therefore non-uniform. Aft end design is a significant factor in shaping this flow.

On a conventional single or twin screw ship the aft-end design is limited by shafting line, propeller clearances and rudder locations. With an RDP the propulsion units can be placed in a location that stops the flow from being impeded (e.g. under the hull). This decreases variations in blade loading and promotes cleaner flow which would be expected to increase the propeller efficiency. This can be matched with either a pram-type stern or bulbous stern, both of which promote buttock flow and decrease overall hull resistance. There is a potential risk that the flow may stall slightly due to the hull not accelerating flow into the propeller; however, the ducting of the RDP would seek to offset this. The duct would increase propulsive efficiency for a given propeller size for high thrust, slow speed applications (e.g. oil tankers), but higher speed vessels (e.g. cargo ships or naval vessels) may be negatively affected by the duct.

There are several usability issues on vessels employing a RDP. First, most designs are confined by draught and therefore the design would need to place the RDPs in a stern tunnel configuration, offsetting many of the gains with regards to clean flow. Second, there still needs to be sufficient flow over a control surface to enable manoeuvrability even at slower speeds. The RDP could azimuth, but previous experience with podded propulsion has shown this to be unreliable (with pods being lost in transit). Finally, any type of unit attached to the vessel with a large projected area will have a significant drag component both from the struts and from the unit itself.

3.2 CURRENT RDP APPLICATIONS

RDPs currently exist as COTS technology and are used as bow/stern thrusters and as propulsion units in azimuth pod and retractable configurations. RDP units are in-service across a range of commercial applications, including yachts, ferries, tug boats, offshore support vessels, research vessels and cruise liners. In total, it is estimated that there are between 100 and 200 rim drives in service within commercial applications. The majority of these units are below 0.5 MW rated power.

To date, propulsion applications have been a smaller part of the market, compared to tunnel thrusters. Frazer-Nash has recently undertaken an investigation into propulsion applications, which involved testing of two 500 kW propulsion units on a research vessel. This trial demonstrated improvements in energy consumption and vibration transmission compared to conventional shaft driven propellers and suggests that the technology offers an attractive solution for future electric ships.

A number of examples of larger scale RDPs are also coming on to the market that will play a role in the propulsion of future platforms. Examples of these include RDPs of up to 1 MW from Schottel [20] and Brunvoll [21] and between 1 MW and 3 MW from Rolls-Royce [19] and Voith [21].

4. TECHNOLOGY SCALING AND INTEGRATION CHALLENGES

4.1 PROPULSION LEVEL RDP POWER RATING

A key constraint in the wider utilisation of RDP systems, particularly for full propulsion applications, is the availability of high power units. The currently available power ratings may be suitable for smaller scale ships (e.g. offshore supply vessels), however there is a disparity between the RDP ratings available currently and the size of the motors used in larger commercial and naval applications. While there are no fundamental issues which would prevent RDP systems being scaled up, significant design changes are likely to be required for the multi-megawatt units needed for application of RDPs on larger vessels. Issues may include the design of bearings and cooling. There will also be significant costs associated with design, manufacture and testing of these higher power devices. Some research has been conducted on highly novel superconducting electrical machine types to offer a more compact and power-dense solution (Hassannia & Darabi, 2014).

4.2 SHIP INTEGRATION

There are several challenges of scaling up this technology to larger ships from a hydrodynamic and structural design point of view. A single large RDP will experience large drag and thrust components and subsequently there will be high cyclic loading on the structure. Podded propulsion units have had issues with being lost in heavy seas and it is envisioned that RDPs could be even more susceptible to this occurring. To minimise this risk the RDPs would either need to be anchored by multiple attachment points or a large single strut: in either case the likelihood is that azimuthing capabilities would be impacted. In addition, single propulsion units would be unsuitable due to risk of loss and so multiple units would be required.

Given that single commercially available units are unlikely to be sufficiently rated to provide all the propulsion power needed for the vessel, the RDPs would need to be stacked. This could be configured advantageously by stacking in a contra-rotating fashion (which is easier to achieve for an RDP given the flexibility in location). While increasing the number of RDPs required, this could potentially provide increased recovery of rotational losses, leading to higher efficiency.

4.3 ELECTRICAL SYSTEM INTEGRATION CHALLENGES

There are two key interrelated issues around the electrical supply to the RDP which have an impact on the scalability of RDP systems: the availability of power electronic drives and the system operating voltage.

A power electronic converter is required to convert the ship electrical power supply (which could be AC or DC depending on the application) to a variable frequency AC supply for the motor. The frequency is usually used to control the speed of the electrical machine.

The electrical machines in RDPs are typically designed with a high number of pole pairs (these being pairs of machine windings and magnets) for optimal size and efficiency (Lai, 2006). This has an impact on the required electrical output frequency of the drive. This frequency can be derived from where f is the required drive output frequency (Hz), S_{RPM} is the required rim drive rotational speed (rpm) and p is the number of pole pairs. The number of pole pairs will likely increase with larger RDP diameters. However, as the rotor speed often decreases as diameter increases in order to limit the tip speed (for performance and noise reasons), the required frequency may not vary widely with changes in design.

$$f = \frac{p.S_{RPM}}{60} \tag{1}$$

Despite the low RDP rotor speed (due it larger diameter), the electrical frequency from the drive is much higher than typical industrial motors of similar power. These typically are designed around 60 Hz operation compared a possible range of 100-300 Hz for RDPs. Ultimately, this requires the drive to be switched faster, increasing switching losses. Managing the associated thermal issues can be more challenging as power levels scale up and this reduces the availability of drives.

Based on a survey of drives conducted by Frazer-Nash, there are a range of COTS options which could broadly meet the power and frequency requirements for relatively low power RDPs (i.e. around 1 MW). These are based on supply from a low voltage (LV) electrical distribution networks (i.e. < 1 kV).

Beyond this point the identification of suitable drive options becomes more challenging. There are two key reasons for this as outlined below.

If supplied from an LV network, the required current rating of the drive switches becomes high as the power of the motor increases. This has implications for the size of the drive as switches are rated on current rather than power. The use of LV distribution for high power propulsion also has wider implications for the rest of the ship electrical system. Issues include: larger and heavier cables, high fault current levels (necessitating uprated circuit breakers) and the potential for high incident energy during fault conditions.

If higher voltage distribution is used to reduce issues on higher power platforms (as it often is, e.g. 4.16 kV is used on the Type 45 [2]), alterative challenges are posed for the drive design. Power electronics switches rated to operate at higher voltages tend to switch more slowly (e.g. < 1 kHz compared to 10 kHz). Therefore despite power output levels being more attainable with higher voltage operation, the required output frequency is more difficult to achieve.

Restrictions on operating voltage can in part be managed using transformers, however this does have size, weight and cost implications for the overall system design.

Alternative solutions include the use of more novel and complex drive topologies such as multi-level or cascaded drives. Figure 4 illustrates one particular example of this type of drive. These multi-level converters can effectively split the system voltage across multiple low voltage switches, enabling fast switching capability to be maintained. Whilst some COTS options do exist for multi-level drives, they are a less mature market and availability is therefore reduced.

4.4 NOISE AND VIBRATION

Noise and vibration produced by the propulsion system is often a significant consideration, and this becomes more pertinent as power levels increase. Rim drive manufacturers suggest that noise and vibration performance is improved over conventional electrical propulsion systems. Whilst rim drives do offer the potential for excellent noise and vibration performance, this capability should not be assumed.

In particular there are some key issues around the acoustic performance of rim drives. This is highlighted within (Carter, 2017), which analyses measurements from a prototype rim drive. A significant difference between RDPs and more conventional propulsion systems is that the motor is directly in the water, rather than inboard where it may be isolated. Therefore any vibrations within the electrical motor are more likely to radiate into the water.

There are a range of sources of noise within the rim drive. These can be broadly categorised as: mechanical, fluid dynamic (hydrodynamic / hydroacoustic) and electromagnetic noise sources (Carter, 2017). Electromagnetic noise associated with the harmonic content of the power supplied to the motor was one potential issue identified in (Carter, 2017). This can lead to high frequency radiated noise associated with the control of the motor and its power electronic drive. This highlights that noise is not solely due to the RDP design but also has implications for the power electronic design and its associated filter. Filtering in particular may add a substantial size and weight burden to the system and is an important consideration within the design.

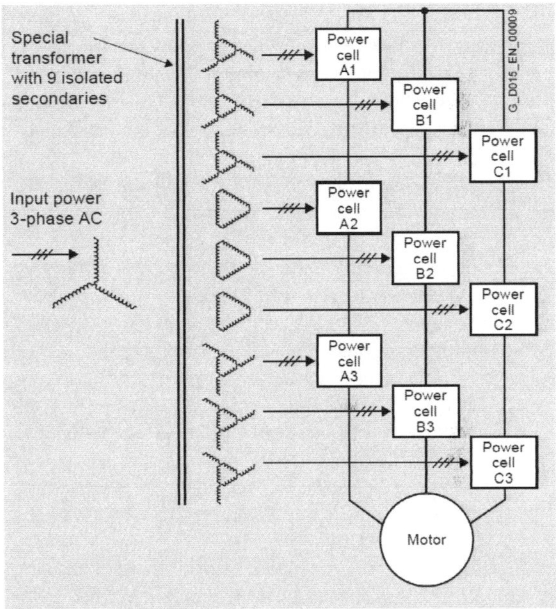

Figure 4: Topology of ROBICON Perfect Harmony drives (3 cells) [25]

This again may lead to novel drive designs being considered for future applications and this is an area of ongoing research. Frazer-Nash are currently working with partners to develop solutions to mitigate these electromagnetic noise issues. These include novel machine, drive, filter designs as well as the using more sophisticated methods of control.

5. CONCLUSIONS

There are a range of potential benefits to using rim drive technology and these are becoming more attainable with an increased range and size of RDP technologies coming onto the market in recent years. Currently available technologies may offer propulsion solutions for smaller ship applications, or as part of a hybrid system with more conventional electrical propulsion options.

While the scaling up of RDP for higher power applications appears achievable, significant technology development and demonstration will be necessary in order to tackle the unique challenges which rim drives present. On this basis, it may be a number of years before RDPs fully challenge conventional electrical technologies for large scale ship propulsion.

6. REFERENCES

1. *Reducing emissions from the shipping sector* https://ec.europa.eu/clima/policies/transport/shipping_en
2. *Type 45 to get third diesel generator to overcome problems*, available online from: https://www.rina.org.uk/Type_45_to_get_third_diesel_generator_to_overcome_problems.html
3. *The Royal Navy and the Next Generation Frigate*, available online from: http://www.gepowerconversion.com/inspire/royal-navy-and-next-generation-frigate
4. *GE Delivers Last of Eight Propulsion Motors for Giant UK Aircraft Carriers* http://maritime-executive.com/article/ge-delivers-last-of-eight-propulsion-motors-for-giant-uk-aircraft-carriers
5. *DDG 1000 Zumwalt Class - Multimission Destroyer*, available online from: http://www.naval-technology.com/projects/dd21/
6. *ABB wins order for Arctic ice-going LNG carriers*, available online from: http://www.abb.com/cawp/seitp202/de3b546300c71b3bc1257ca60048d1d1.aspx
7. *Yamal LNG (Russia)*, available online from: http://sovcomflot.ru/en/fleet/business_scope/projects/item1658.html
8. *Elation - Cruise Liner*, available online from: http://www.ship-technology.com/projects/elation/
9. L3, *Diesel-Electric Propulsion Systems*, http://www.l-3mps.com/products/datasheet/SAM/1.002.pdf
10. Siemens, "*Reference List - Offshore Vessels - Low Voltage Propulsion*", https://w3.siemens.no/home/no/no/sector/industry/marine/Documents/Offshore-References%20V2012_02.pdf
11. *Multi Maritime*, available online from: http://www.multi-maritime.no/two-lng-powered-ferries-of-mm-design-to-be-built-by-vard/
12. Siemens, "*Setting a Course for Carbon-Free Shipping*", available online from: https://www.siemens.com/innovation/en/home/pictures-of-the-future/mobility-and-motors/electromobility-electric-ferries.html
13. Rolls-Royce, "*Hybrid and electric propulsion systems*", available online from: https://www.rolls-royce.com/~/media/Files/R/Rolls-Royce/documents/customers/marine/hybrid-electric-propulsion-systems.pdf
14. HOORN A., *et al* "*Contra-rotating propellers*", available online from: http://www.maritimesymposium-rotterdam.nl/uploads/Route/Contra-rotating%20propellers.pdf
15. LU, R,. TURAN, O,. BOULOUGOURIS, E., BANKS, C., INCECIK, A., "*A semi-empirical ship operational performance prediction model for voyage optimization towards energy efficient shipping*", In Ocean Engineering, Volume 110, Part B, 2015, Pages 18-28, ISSN 0029-8018,
16. VAN BLARCOM, B,. HANHINEN, J,. and MEWIS, F,. "*The Commercial Rim-Driven Permanent Magnet Motor Propulsor Pod*" available online from: http://www.friedrich-mewis.de/files/Mewis/Publications/22.pdf
17. CARTER, C,. "*Rim drive propulsion in the underwater defence sector; applications, benefits and challenges*" Undersea Defence Technology Conference, May 2017
18. HAMSON K,. and WARD, C,. "*Underwater Radiated Noise from Electrical Propulsion*", Undersea Defence Technology Conference, May 2017
19. Rolls-Royce, *Azimuthing permanent magnet thruster*, "http://www.rolls-royce.com/~/media/Files/R/Rolls-Royce/documents/marine-product-finder/rr-az-pm-broch190816.pdf
20. Schottel, "*System Solutions – Propulsion And Manoeuvring*", https://www.schottel.de/marine-propulsion/
21. Brunvoll, *Rim Driven Thrusters*, http://www.brunvoll.no/product/the-complete-thruster-system/rim-driven-thrusters-rdt
22. Voith, "*Voith Inline Thruster / Voith Inline Propulsor*," available online from: http://www.voith.com/en/products-services/power-transmission/voith-inline-thruster-voith-inline-propeller-11014.html
23. HASSANNIA A,. and DARABI, A ,. "*Design and Performance Analysis of Superconducting Rim-Driven Synchronous Motors for Marine Propulsion*," in IEEE Trans. on Applied Superconductivity, vol. 24, no. 1, pp. 40-46, Feb. 2014.
24. LAI S.H., "*Design Optimisation of a Slotless Brushless Permanent Magnet DC Motor with Helically-Wound Laminations for Underwater Rim-Driven Thrusters*" PhD thesis, March 2006
25. Siemens, "*Medium-Voltage Liquid-Cooled Drives*", Catalog D 15.1, 2012, available online from: https://www.industry.usa.siemens.com/drives/us/en/electric-drives/medium-voltage-drives/Documents/DRV-MV-liquid-cooled-catalog.pdf

7. AUTHORS BIOGRAPHY

Steven Fletcher is a Senior Engineer in the Technology Management group within Frazer-Nash Consultancy. He gained his PhD from the University of Strathclyde in the field of electrical power systems. He specialises in the design and modelling of electrical systems for transport applications.

Robert Hayes is an Engineer in the Marine Technology group within Frazer-Nash Consultancy. He has a degree in Ship Science from the University of Southampton. He is a naval architect and specialises in the assessment of hybrid and novel marine propulsion technology integration.

Callum Morgan is a Senior Engineer in the Electrical, Instrumentation and Control group within Frazer-Nash Consultancy. He has a degree in Electrical and Mechanical Engineering from the University of Edinburgh. Specialises in electrical system design and safety analysis.

Andrew Allan is a Consultant at Frazer-Nash Consultancy with a background in Electrical and Mechanical Engineering. He specialises in mathematical modelling and simulation of multi-domain systems and has considerable experience modelling naval electrical power and propulsion systems.

Chris Carter is the Underwater Technology Business Manager at Frazer-Nash and oversees Frazer-Nash's underwater research and technology activities. Chris is a Chartered Engineer and an experienced project and programme manager with a technical background in fluid dynamics, structural integrity, signatures and technology management.

REDUCE PORT AND SHIPPING EMISSIONS – THREE STEP PROCESS

B Perry, PBES, Norway

SUMMARY

Reducing port and shipping emission – The three step process

- Zero Emissions in Port – Tugboats and other support craft
- Coastal Shipping and Ferry Hybridization/Electrification
- Shore Power – how batteries reduce burden on local grid and expand availability

1. INTRODUCTION

Today, more than 100,000 vessels travel through 4,500 ports worldwide, produce emissions equivalent to 220 coal-fired power plants. (Zolciak, 2016)

With seaborne trade expected to double to as many as 24 billion tons annually come 2030, (ABB, 2016) the shipping industry must reduce its carbon footprint and adopt technologies that maintain ability to economically conduct trade.

Now more than ever it is apparent that the marine industry must act on climate change. Recent global climate events underscore the need to shift away from traditional fossil fuel as a power source. New national and international regulations are forcing operators, owners and shipbuilders to look at alternate sources of propulsion energy, the most effective of which is energy storage in the form of lithium batteries used to hybridize and fully electrify ships.

People are just beginning to consider battery technology to support clean port operations, and may be hanging back assuming the business case isn't yet set. However not only is it technologically possible to replace diesel engines with battery-powered electric motors but according to a Siemens study, 30 ferry routes in Denmark alone, would be more profitable with electric propulsion compared to diesel. This feasibility study investigated 52 of Denmark's coastal passenger ferries that carry inhabitants, cars and visitors on 42 domestic routes, looking at the potential for replacing Denmark's traditional diesel-powered ferries with new electrical ferries that are both greener, and also have the potential for delivering a stronger business case. (Siemens, 2016)

2. ZERO EMISSIONS IN PORT

The roadmap towards a transition to a low-carbon society has begun, and already we are seeing what it will take to meet climate goals. A DNV GL study shows that even if all known maritime energy efficiency techniques are employed, it will not be possible to reduce emissions from domestic shipping by 40%, government's targets for emissions reductions will not be met with energy efficiency measures alone. While energy efficiency techniques such as speed reduction, propeller polishing or hydrodynamic improvements are helping to reduce ship emissions, they are limited in effectiveness. In order to achieve the goals set out by the EU, battery power is required.

Zero or low emission harbour activities are now available to a variety of vessels and machines in ports. Each type utilizes battery power. The presentation will explore how various vessels and machinery can implement the technology today with economic benefit.

3. TUGS

The general increase in vessel size, such as that experienced within the container and tanker sectors, has pushed up the scale and power of the tugs needed to handle them. Tugs tend to be defined by their peak bollard pull, while at the same time, the nature of harbour operations demand that engines possess adaptability in low-load operations. The issue has always been that harbour operations require long periods on standby then a sudden demand for full power. So, there's a mismatch that's only going to grow.

Figure 1 is a useful outline of a typical harbour tug profile – note that demands over 60% loads only accounts for around 3% of the total time.

Figure 1 (Boyd, 2014)

Mode-ID	TASK	SERVICE	SPEED-(knots)	TIME-(%)
1	Stand-by	Idle	0	15%
2	Transit-low	Transit	6.6	30%
3	Transit-high	Transit	10	7%
4	Assist-80%	Tow-pull-[80%]	1	1%
5	Assist-60%	Tow-pull-[60%]	1	1%
6	Assist-40%	Tow-pull-[40%]	1	9%
7	Assist-20%	Tow-pull-[20%]	1	26%
8	Barge-move-60%	Tow-pull-[60%]	5	1%
9	Barge-move-40%	Tow-pull-[40%]	5	1%
10	Barge-move-20%	Tow-pull-[20%]	5	9%

More, black smoke near to shore, typical of very low loads, is never good. As landside communities grow and become more aware of the impact of NOx and other particulates and the supply chain more sensitive to the eco-agenda, the bar will inevitably be raised.

"We believe that new tugs will in future rely more and more on battery and hybrid propulsion," said Riku-Pekka Hägg, Vice President of Ship Design at Wärtsilä Marine Solutions. The first 75 tonne bollard pull tug with a HY hybrid power module is due to begin operating in the Port of Genoa for the biggest tug operator in Italy, Rimorchiatori Riuniti, early next year.

4. SUPPORT BOATS

However, while in some areas the support boat market is suffering from the economic downturn arising from the collapse in oil prices, in others, continuing port expansion and shipyard development is generating demand for various types of support vessels.

Despite the balancing action on upfront costs as a hybrid layout can reduce the size of the engines and other components, there's occasionally still some reluctance to invest. This is partly because quite often fuel-saving benefits often accrue to charterers who pay for fuel – so there's a disconnect between those investing in a better, greener vessel and those footing the bunkering bill.

However, there are signs this issue might be about to change, as in Norway, some new charter arrangements have been developed for the offshore oil and gas market.

The last few months has seen Statoil fix charters for several offshore vessels that will be upgraded with hybrid battery propulsion as a condition of their contracts. All of the vessels in question will be equipped with hybrid battery operation, with the possibility for shore power connection, also allowing them to reduce fuel consumption especially while working in dynamic positioning mode.

There's more to it than the fuel bill: the so-called spinning reserve can be replaced now with battery systems making for lower maintenance patterns, batteries also smooth the load by compensating for peaks and troughs as well as enhancing safety and reliability by providing back-up to mitigate against blackouts – a situation problematic for OSVs that need to hold station close to large offshore structures.

5. WIND SERVICE VESSELS

Recently, a new Louis Dreyfus Armateurs wind farm Service Operation Vessel (SOV) is looking at integrating batteries. This represents a new sector which is implementing the use of energy storage technology. This might even tie in with a new type of windfarm operation pattern. As some windfarms are pushing further offshore, making daily trips less feasible. While till now CTVs have normally been fast paced vessels with engines tailored for long transits, if a mothership pattern is established instead, it may be that the smaller electrical transfer boats could be used, picking up a battery charge while linked to the larger vessel.

6. OTHER SUPPORT CRAFT

Next year will see the delivery of the Norwegian Coastal Administration's new multipurpose vessel OV Ryvingen, its second to combine traditional diesel power with battery-driven operation. The battery pack will have an available capacity of 2,000 kWh – that is a substantial increase from the 850 kWh capacity installed aboard its hybrid sister ship the Bøkfjord.

The Ryvingen has an operating profile that allows it to run on battery power alone for several hours, without using the diesel engine. In addition, the batteries will provide power when the vessel is docked, so the diesel engines will not have to be kept running. The batteries can be recharged from an onshore power supply in ports where this is available.

7. COASTAL SHIPPING

Coastal and inland shipping are both prime for hybrid propulsion, especially as battery technology has increased in effectiveness and cost has decreased. Barges and ships can utilize plug-in hybrid propulsion as prime movers to reduce costs and emissions, in an environment that is actively looking to enact a modal shift, taking cargo from highly polluting road trucks to waterways.

Some regions are already taking the challenge seriously. In Norway in 2015, DNV GL took the lead in creating the Green Coastal Shipping Programme. Last year programme director Narve Mjøs predicted: "We will see developments; similar to those we are now seeing in the automotive sector, in which plug-in hybrid (rechargeable) ships will also be an integral part of the land-based power system. This September it has been underpinned by a report (compiled under the ReCharge initiative), by Cavotec, ABB, DNV GL and the Port of Oslo, which has identified 64 vessels with an operational profile suitable for battery propulsion, and outlined charging infrastructure on 97 different routes to support it. This would lead to a possible 12,517 tonne reduction of CO_2 and 13,133 kg reduction of NOx emissions. (Cavotec, 2017)

8. FERRIES

Small, short hop ferries (both passenger and ro-ro) usually face a disproportionate amount of time manoeuvring out of berth then accelerating, barely touching full speed before decelerating. It's simply not a good use of diesel engines, especially since the quays are often near or actually inside built up areas. Many countries, such as Norway, are thinking in terms of totally electric running rather than hybrids, as these systems are simpler and the most effective at targeting emission reductions. The weight is often not an issue and the batteries can take the place of ballast.

Bigger ferries are also starting to pick up hybrid and electric technology. Norway's Color Line signed a letter of intent with Ulstein Verft to build a 160m long vessel capable of carrying 2,000 passengers and 500 cars on a route between Sandefjord and Strømstad. The ferry will switch to electric-only power in Sandefjord and when docking at both ends of the route. The battery, which can power the ferry for about half an hour, can be recharged on the dock and by the diesel engine. More, HH Ferries Group's Tycho Brahe and the Aurora, operating a 4km route between Helsingborg (Sweden) and Helsingör (Denmark) are to be converted to full electric running: these ferries show the scale is extending: these are 238m long and weigh 8,414 tonnes, carrying 7.4 million passengers and 1.9 million vehicles annually.

8.1 HIGH SPEED FERRIES

Full electrification of high speed ferries, though, have hit a snag: weight. The amount of energy needed to propel the boat forward increases faster than the speed, so that before long, you are using most of the power to push the batteries. However, both Norway and the US' Sandia National Laboratories are now beginning to look at putting together hydrogen and batteries: a natural pairing especially as fuel cell development is maturing rapidly.

However, as noted electric ferries often require a fast charge at either end of the route, and batteries, again, play an important role.

9. SHORE POWER

Traditional shore power connects directly to the grid; the addition of rechargeable battery banks (containers full of batteries) expands availability. This could also answer the 'fast charge' demand of these shorter run ferries.

There is also the possibility of using off peak or renewable power for port handling equipment, freeing facilities from the limitations of power availability or infrastructure capacity. Battery packs are charged by grid or other renewable sources, and can be used to optimize and reduce on-shore power generation or provide emergency spinning reserve and UPS for the port and surrounding areas.

10. REMOTE AND AUTONOMOUS RUNNING

PBES monitors its systems 24/7. If we see a slight variation in voltage, we know it before the customer does. We are very proactive in maintenance and service and support, and so we can give a much better warranty because we always have eyes on the system. This energy storage system is a wireless hub. This is the Internet of Things. We can dial into these from anywhere in the world and fix something, change programming, update software. We don't have to be on the boat.

However, in the last few months there's been a new development, as one-way, remote monitoring has changed into a two-way street: a massive cargo ship off the coast of Scotland has been successfully piloted – from San Diego, California. It won't end here: full autonomy is just around the corner for some segments.

For example, the autonomous Yara Birkeland, a 120 teu coastal feeder, will sail between three ports in southern Norway: Herøya, Brevik and Larvik. It is going to take around 40,000 journeys per year from the road. Development of coastal vessels like Birkeland will have an advantage as they don't have to wait for IMO regulation, but will rely on the willingness of the flag state to move its own regulation – as has been seen in Norway.

It makes sense that the change to autonomous vessels will come with electric power, rather than diesel-electric as controls are simpler and more effective. More, autonomous vessels get better battery performance than humans as they reduce "wastage" of power through predictive responses that require fewer corrections.

But the revolution to take the drudgery out of harbour monitoring and standby operations is already underway and has been for some time, there are numbers of small underwater and surface craft near finalisation – take Eco Marine Power's Aquarius Unmanned Surface Vessel - that can patrol a harbour, relying on electric power and in some cases, photovoltaic charging. More, now that drones are now in everybody's pocket, maybe at some point we'll even see ports incorporate a recharging hub for operational inspection.

For those asking questions about the real impact of batteries, a recent Norwegian study took a hard look at the true environmental costs and benefits by examining two ship types: a platform supply vessel (PSV) and a ferry. The environmental costs and benefits, calculated as global warming potential (GWP) and NOx emissions, show that the environmental CAPEX of producing batteries for a PSV and ferry is negligible compared to the fuel savings from using batteries. In fact, although there are some aspects of uncertainty regarding the environmental CAPEX of the battery systems, these uncertainties are not large enough to change the conclusion. Even if the environmental CAPEX increases by a factor of 10, the conclusion will be not change for an OSV or for a ferry employing Norwegian electricity mix. (Henningsgard, 2016)

11. CONCLUSION

Bringing the electric revolution to the seas has reduced emissions but is also proving to be financially profitable to ship owners. More, with the growing green agenda, it's providing a way for ship owners, which after all, buy ships with a design life of two or three decades, to 'future-proof' operations against the rising tide of regulation and even lead public opinion.

The use of energy storage to provide power for emission free operation is here today. Using this industrial evolution of propulsion, the marine industry is able to lower operational costs and provide meaningful return on investment all while doing its part, and moreover, show it's willing to take responsible, leading action, in the fight against climate change.

12. REFERENCES

1. ZOLCIAK, A., '*Running the world on clean energy requires innovation*', 2016.
2. ABB, '*ABB creates vision of future – starting today – with 'Zero Emissions Port Call*', June 2016.
3. SIEMENS, '*Electrification of Denmark's Ferry Fleet*', October 2016.
4. BOYD, E., '*Using Detailed Vessel Operating Data to Identify Energy-Saving Strategies*', 2014.
5. CAVOTEC, '*ReCharge*', September 2017.
6. HENNINGSGARD, S., 'Life cycle assessment of batteries in maritime sector', *Maritime Battery Forum*, September 2016.

13. AUTHORS BIOGRAPHY

Brent Perry is one of the most knowledgeable and experienced people in the marine industry in regards to practical use of energy storage in commercial applications. His 30-year history in shipbuilding and deep knowledge of large-scale energy systems give him a unique perspective on the industry. An accomplished, engaging and passionate speaker, Mr. Perry resides in Vancouver, British Columbia and enjoys sailing with his wife and three children.

Brent Perry holds the current position of CEO at PBES. His previous experience includes 30 years in the marine industry.

REPLACEMENT OF A DIESEL GENERATOR WITH A CONTAINERISED BATTERY SYSTEM ON-BOARD A CONTAINERSHIP

I Dimakopoulos, **P Menegakis**, **N Lampris**, **S Gkinis** and **G Panagoulias**, Newcastle University, UK

SUMMARY

The outcomes from the two IMO Greenhouse Gas Studies (2009 & 2014) (IMO, 2008), (IMO, 2016), highlighted the magnitude of shipping's impact on climate change. This, in combination with the rather stringent regulations, which are continuously being adopted, have resulted in the development of innovative and energy efficient technologies aiming at mitigating shipping's impact on the environment. This design project proposes an energy efficient solution aiming at the reduction of air emissions generated by the fleet of one of the largest sections of international shipping, containerships. By designing a containerised battery system capable of providing 10.3MWh of useful energy, the ultimate aim is to replace an existing diesel generator and to operate the battery system on manoeuvring. With the aid of a load planner software it was ensured that such a system can be loaded on-board a specific containership, without a negative impact on the stability and the structural integrity of the employed vessel. The battery system can be charged both by a shore supply (cold ironing) as well as during the seagoing part of the voyage. The energy feasibility of such a system and the emissions reduction were verified by developing several software models, which allowed various case study trips to be investigated. Results from the case studies suggested that with such a system on-board, emissions can be reduced by 19% when the ship carries only normal containers and by 4 % when the ship carries normal and refrigerated containers. However, the performed financial analysis indicated that such an investment would result to a loss, primarily due to the current price of battery systems.

1. INTRODUCTION

The need of greener ships combined with the increased pressure of complying with IMO legislation have created high expectations in international maritime sector. The scope of this project is to present a novel idea addressing the environmental demands of the future by designing a battery system, the energy of which will be used during the manoeuvring operation of the vessel. The battery system will be charged both by a shore supply (cold ironing) as well as from the auxiliary engines currently on-board (operating them at their most efficient point).

1.1 ENVIRONMENTAL NEED

The necessity of additional administrative actions is outcome of environmental and economic incremental pressure on shipping. Maritime industry is estimated to produce 3% of greenhouse gas emissions worldwide which is predicted to increase further due to the growth of world trade (Doo & Lee, 2013). Dealing with this crucial issue has become a legal and industrial requirement as well as a society's expectation.

Within the shipping sector a variety of distinct operational and technological options have been presented to potentially assist in reaching reduction of emissions targets. In the last decades, shipping's impact on climate change has been addressed by the International Maritime Organisation through conventions and international regulations. Therefore, amendments which are related with Annex VI of IMO were introduced aiming to adequately reduce NO_x and SO_x emissions (IMO, 2008).

Shipping sector has a significant part worldwide as it facilitates trade among nations. On the other hand, it has critical contribution to climate change (by producing greenhouse gases) and this is the reason of establishing several initiatives which have been adopted over the past years. However, there is still room for improvement regarding the mitigation of greenhouse gas emissions in the shipping industry.

1.2 AUXILIARIES AND MANOEUVRING

The objective of this paper is to propose an alternative idea which aims to reduce the percentage of GHG emitted as a result of auxiliary engine usage. Auxiliary engines are vital for the wellbeing and operation of a vessel but simultaneously have a decent role to the environmental pollution augmentation. The diesel generator which constitutes part of the auxiliary engine system, approximately accounts for 28% of the total fuel oil consumption on-board. Thus, the diesel generator is responsible for 28% of the total emissions produced by a vessel (Parker, *et al*, 2015). According to these percentages, it can be deduced that auxiliary engines have a significant effect on ship's total emissions and as a result, there should be a scope of reducing these percentages.

The auxiliary engines are being used during all operations of the vessel (seagoing and manoeuvring). However, their emissions contribution on seagoing and manoeuvring is huge in contrast with their role as a compartment of a vessel. As an indication, emissions during manoeuvring operation are at a level of 18.1% of the total emissions during a vessel's trip from the departure port to the arrival port. As far as containership's NO_X emissions during manoeuvring operation are concerned, 64% is coming from the auxiliary engines of the vessel (Kevin, *et al*, 2006).

Hence, these results indicate an urgent need for the reduction of the auxiliary engines' emissions both during the seagoing part and the manoeuvring operation of a given voyage. This paper aims to reduce these emissions by investigating the possibility of replacing one diesel generator with a containerised battery system on-board a containership.

2. BATTERY SYSTEM

2.1 BATTERY SELECTION

2.1.1 Energy Demand Identification

The following picture demonstrates the vessel (*"IRENES REMEDY"*) (Anonymous, 2017a) on which this project emphasises. It is a feedermax containership (IMO number: 9315850) with a total capacity of 2,824TEU containers.

Figure. 1. IRENES REMEDY (Marinetraffic.com, 2017)

The aim of this paper is to use the stored energy in the battery during manoeuvring of the vessel in order to cover its auxiliary power demands. The battery system would be the only energy supplier when the ship carries only normal containers (non-refrigerated) for a manoeuvring period of up to 8 hours. As for the scenario of the ship carrying refrigerated containers both the battery system and the auxiliary engines would be energy suppliers. However, under this scenario the running auxiliary engines would be less in comparison with the current power configuration.

The first step in the design of the battery system was to make use of the voyage summary of the vessel. Analysis of the voyage summary showed that 95% of the manoeuvring times were less than 8 hours. Hence, it was decided to design the battery system to be able to provide the necessary energy for these 8 hours of manoeuvring when the ship carries only normal containers and no refrigerated ones. Afterwards, the electric load analysis of the employed vessel was utilised and as it was advised by the chief engineer on-board, during a private communication, for a manoeuvring period of 8 hours the hoteling loads for the first 7 hours would be 1,143.8 kW and for the final 1 hour the load would be 2,301.7kW. As a result, the energy required to cover vessel's auxiliary demands for 8 hours of manoeuvring would be 10.3MWh.

2.1.2 Battery System Design

It was decided to design the battery system with products offered by CORVUS Energy Inc., which is a leading lithium-ion battery manufacturer for marine applications and its products are approved by several classification societies. The battery system would consist of battery packs with the following technical specifications.

Table 1. Battery pack's technical specifications (corvusenergy.com, 2017)

Quantity	Value	Quantity	Value
Energy (kWh)	125	Width (m)	0.87
Max. Voltage (VDC)	1100	Depth (m)	0.71
Nom. Voltage (VDC)	980	Weight (kg)	1550
Min. Voltage (VDC)	870	Depth of Discharge (DoD)	0.8
Height (m)	2.2		

The connection configuration of the whole system will consist of the following equipment: step-up/step-down transformer, inverter/rectifier unit and cables. Taking into account the efficiency of those systems, the efficiency of the battery packs as well as the DoD of the packs it was deduced that the battery capacity should be 15.0MWh. This capacity can be accomplished by placing 120 battery packs in parallel.

2.1.3 Battery System Storage

It was decided to place the battery packs inside TEU refrigerated containers (5.43m long and 2.28m wide) together with the Battery Management System (BMS). After a private conversation with CORVUS Energy Inc., it was advised that within the container half of the space is devoted to the battery packs and the other half to the BMS.

Analysis showed that 6 battery packs can fit within a single TEU refrigerated container. Hence 20 such containers will be required for the entire battery system, with each one having an approximate weight of 20tonnes.

2.2 BATTERY ARRANGEMENT

The type of the container that was decided to be used is a 20-ft. refrigerated container. Due to the nature of the containership there is infrastructure compatibility as in the container seats there is access to the ship's electrical grid to provide energy for the refrigerating system. As analysed earlier the dimensions of this type of container are ideal for six battery packs and the components that support them. The refrigerating ability of this container type enables the control of the operating environmental conditions such as temperature and humidity, as recommended by the battery manufacturer.

The arrangement of the battery system within each container unit was generated as a 3D model using Autodesk Inventor. A screenshot of this model is shown below in Figure. 2:

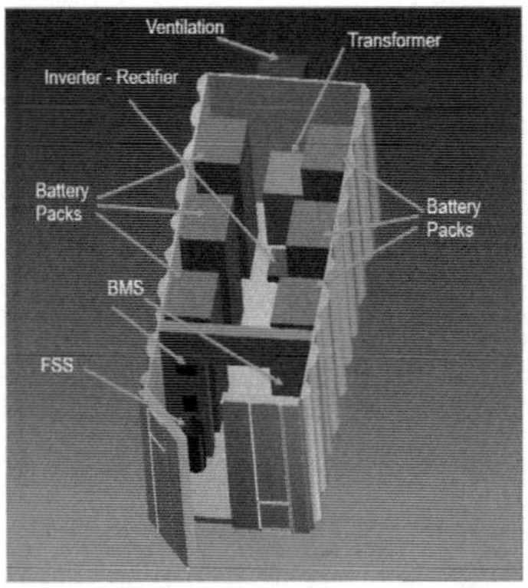

Figure. 2. Containerised Battery - 3D model

The above figure demonstrates the arrangement within each container unit. The six battery packs are being managed by the BMS. The BMS is placed near the entrance of the container to be easily accessible and monitored. Other auxiliary components of the battery system such as the transformer and the inverter/rectifier unit are placed inside the battery compartment. The whole system is supported by a fire suppression system (FSS), which is placed near the entrance, and a ventilation system to refresh the toxic air in the event of ignition of one of the packs.

The physical arrangement of such a system was decided according to the dimensions given by the components' manufacturers and all the necessary equipment is present to support the battery system during different scenarios.

3. CONTAINERISED BATTERY ON-BOARD ARRANGEMENT

As it was mentioned in *Chapter 2,* the proposed battery system (using CORVUS Energy Inc.) (Corvusenergy.com, 2017) was calculated to consist of 20 TEU containers. The total weight of each TEU battery container unit, accounting for the battery packs needed to provide the required energy as well as the various other components of the system (e.g. Battery Management System, Fire Suppression System, Transformer, Inverter/Rectifier unit etc.), was calculated to be approximately 20MT. Additionally, these 20 TEU containers were assumed to be refrigerated type (reefer containers), in order for the battery configuration's HVAC (Heating, Ventilation and Air Conditioning) requirements to be accommodated and to maintain an optimum temperature within the container. Since for the purpose of this design analysis the fitting of 20 TEU containers with the above mentioned characteristics on-board the employed containership was considered, a stability analysis needed to be performed to investigate the optimum place for the battery system.

In order for the optimum battery configuration arrangement to be investigated, the capacity plan of the employed vessel was utilised. However, with a view to identify the most suitable place for the battery container units, some basic information on container numbering should be stated first. The exact position of the containerised battery system on a containership was identified, by utilising the three dimensional system (bay-row-tier). The employed vessel is divided into compartments (bays), which proceed from 01 to 50, where bay 01 is the first compartment at the bow of the vessel and bay 50 is the last at the stern of the vessel. The bays represent the cross sections of the ship and as mentioned above, are numbered from bow to stern. The rows as pictured in Figure. 3 (11-01 and 02-12, "ROW NO.") start at the midship of the ship and increase to both sides, whereas the tiers are layers of containers represented by the following numbers (02-12 and 82-94, "TIER NO.") in Figure. 3 which are numbered from the bottom upwards.

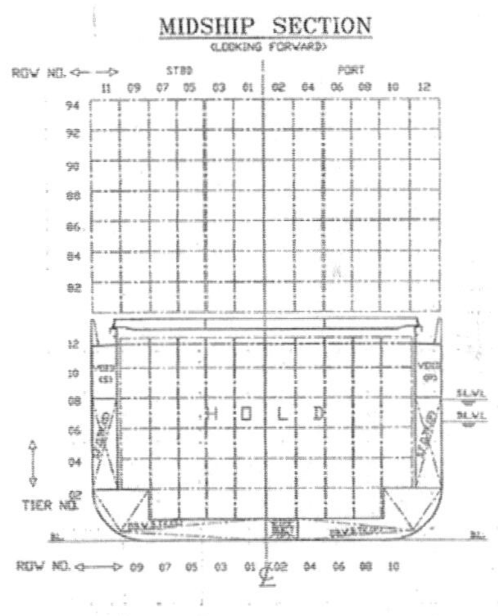

Figure. 3. Midship section - IRENES REMEDY

3.1 OPTIMUM BATTERY SYSTEM LOCATION

After investigating all the places that these 20 TEU container units could be possibly loaded, it was decided that the optimum place would be at the first two tiers of the bay behind the accommodation (aft part of the vessel). In fact, the battery configuration arrangement was decided to be as follows: 2 rows of 10 battery container units, placed on top of each other at tiers 82 and 84 of bay 45. Therefore, the three dimensional system identification numbers of the container units are the following: (from 45-09-82 to 45-01-82, from 45-02-82 to 45-10-82 and from 45-09-84 to 45-01-84, from 45-02-84 to 45-10-84).

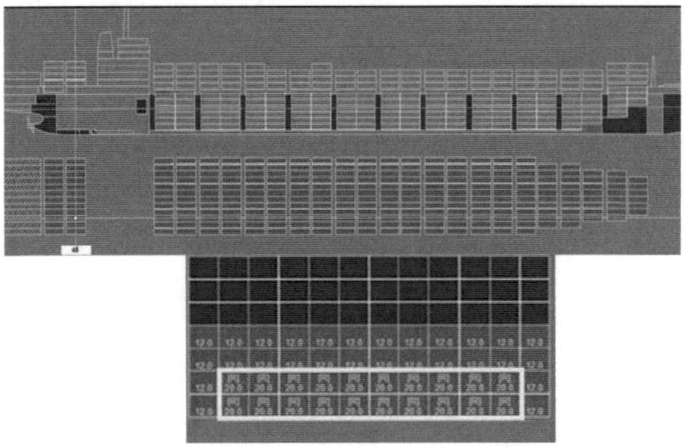

Figure. 4. Battery container units' location (yellow rectangle)

These specific container slots were selected for two main reasons. Firstly, the above identified container slots have the mark (R10), which indicates that there are specific places (reefer sockets, hence the letter 'R') to "plug" the cables from the reefer containers. Additionally, two normal container units can be placed at each side of the 10 slot battery container units, to account for minimum weather protection. In general, a clearer illustration of the optimum battery configuration arrangement can be observed in Figure. 4 (yellow frame).

3.2 IRENES REMEDY LOADING CASE STUDY

In order to investigate whether such a system can actually be loaded on-board a containership, "ANKO" marine load planner software was utilised (Anonymous, 2017b). With the use of this software, a typical loading scenario case study was carried out accounting for the battery system. In this particular case study, a total number of 2,824 containers (total capacity of the vessel) were used (2,804 normal and 20 refrigerated containers which represent the battery container units). An average weight of 12 MT and 17.5 MT was assumed for the TEU and FEU containers (non-battery container units) respectively. It should also be noted, that it was assumed that each stack of containers should not exceed the total weight of 80MT and that only two normal containers are to be placed above the battery container units to prevent excessive weight forces on the battery containers. Additionally, in order for the case study scenario to be as accurate as possible, typical short trip ballast and fuel load conditions were also assumed. Finally, the containers were arranged in such a way that the visibility status of the case study loading condition to be in compliance with SOLAS and Panama Visibility Lines.

3.3 STABILITY ANALYSIS

By simulating the above analysed loading case study scenario, the load planner software provided data regarding the stability of the vessel as well as the bending moments and the shear forces acting on the ship. As it can be seen in Figure. 5, the obtained stability values were found to be within the limits suggesting that the proposed battery system would not have any negative effects on the stability of the employed vessel. Finally, the shear forces and the bending moments acting upon all the frames of the vessel, were again found to be within the limits, since the percentages of the actual values of shear forces and bending moments were calculated to be well below the limit. In conclusion, the case study indicated that the proposed battery container configuration can be combined with a typical container loading scenario without any stability or any other kind of issues (e.g. structural integrity of the vessel to be questionable), suggesting that loading 20 TEU refrigerated containers filled with battery modules and power electronics on-board the employed containership would be possible.

INTACT STABILITY ANALYSIS Sea Density : 1.0250 MT/m3

ITEMS	Weight MT	VCG m	LCG m	TCG m	FSM/HM MT-m
LightShip	11927	12.05	-19.65	0.06	0
Constants	162	17.85	-13.08	0.16	0
Cargo	24597	17.17	-5.11	0.00	0
Ballast	8941	3.23	14.26	0.08	194
Fresh Water	0	0.00	0.00	0.00	0
Fuel Oil	641	6.41	0.02	0.00	65
Diesel Oil	0	0.00	0.00	0.00	0
Lub Oil	0	0.00	0.00	0.00	0
Stores	0	0.00	0.00	0.00	0
Deadweight	34340	13.34	-0.01	0.02	259
TOTALS	46267	13.01	-5.07	0.03	259

STABILITY CRITERIA

Criterion		Actual		Limit	
Area 0-30 deg	m x RAD	0.1378	>=	0.055	OK
Area 0-40 deg	m x RAD	0.2352	>=	0.09	OK
Area 30-40 deg	m x RAD	0.0974	>=	0.03	OK
GZ at/or> 30 deg	m	0.533	>=	0.2	OK
Max GZ Angle	Deg	36.2	>=	25.0	OK
Maximum GZ	0.57 m				
Initial GM	m	1.009	>=	0.15	OK
Weather Area(B-A)	m x RAD	0.1845	>=	0.0	OK
Lim KG	m	13.016		Undefined	NA
Req GM	m	1.009		Undefined	NA
Min FPP Draft	m	11.043	>=	4.3	OK

HYDROSTATICS		
Draft FPP	11.04	m
Mk F106.8	11.05	m
APP	11.16	m
Mk F16.5	11.15	m
Mid	11.10	m
Mk	11.10	m
LCF	11.11	m
TRIM	0.11	m
HEEL	2.0	Deg
LCF	-14.59	m
Prop Imm	101.1	%
Rolling	21	sec
TPC-I	55.17	MT/cm
MCT	756.9	MT-m/cm
MCH	815	MT-m/deg
FLood	53.6	Deg
LCB	-5.08	m
KM(T)	14.03	m
KG	13.01	m
GM	1.01	m
GGo	0.01	m
GoM	1.01	m
KG(eff)	13.02	m

Figure. 5. Case study's stability data

4. DEVELOPED MODELS

4.1 REEFER OPTIMISATION MODEL

In *Chapter 2.1* it has been ensured that the designed battery system is capable of providing the required energy for 8 hours of manoeuvring when the containership carriers only normal containers and no refrigerated ones. Hence when the ship carries normal and refrigerated containers (increased power demand) the battery system cannot achieve meeting the energy demand for 8 hours of manoeuvring on its own. Apart from this limitation, the three remaining generators on-board would not achieve meeting the seagoing power demand and at the same time charge the battery (even if working at 100% loading), when the ship accommodates maximum number of refrigerated containers. Hence the number of reefer containers that can be accommodated on-board needs to be optimised.

Utilising vessel's voyage summary it was obtained that the mean passage time per voyage is 97 hours. Therefore, these 97 hours were used as an input for passage time to the developed model. As far as the auxiliary engines working are concerned, it was decided that 3 engines would be running on seagoing at 85% loading and 2 engines at the same loading on manoeuvring (for a period of 8 hours). As for the number of TEU and FEU refrigerated containers, these have not been considered at this stage because they were the changing variables.

In order to find out the maximum number of refrigerated TEU and FEU containers two prerequisites have been set which must be satisfied simultaneously. The first one is that the energy extracted from the working auxiliary engines on seagoing should be 15.0MWh more than the energy demand of the vessel on seagoing. The second prerequisite is associated with the energy balance during manoeuvring. The energy provided by the running auxiliary engines and the battery system should be equal or more than the energy demand on manoeuvring.

Initially the vessel could accommodate 182 TEU and 404 FEU refrigerated containers and once the containerised battery system would be adopted 53 TEU and 310 FEU refrigerated containers can be accommodated. At this point, it should be noted that the reduced number of refrigerated containers does not mean lost freight for the ship-owner as the "lost" refrigerated containers would be transformed into normal containers. Hence, the payload of the vessel would not change. It should also be noted that from the 53 TEU refrigerated containers that can be accommodated on-board 20 of them would be devoted to the battery system and the remaining 33 would be for transporting cargo.

4.2 EMISSIONS MODEL

A model has been developed which calculates emissions for the auxiliary engines of the vessel for manoeuvring out, seagoing and manoeuvring in, given certain inputs from the user. This model is linked with the Reefer Optimisation Model, explained in *Chapter 4.1*, in the sense that the user enters the number of auxiliary engines running during manoeuvring out, seagoing and manoeuvring in as well as the corresponding percentage loading on the Reefer Optimisation Model. Based on these inputs and on data that were collected from the shop test for the auxiliary engines currently on-board, the model calculates emissions for the three operations of the vessel. The model also takes into account the fact that for a given journey the vessel might sail inside and outside an Emission Control Area (ECA) zone.

4.3 TRIP SIMULATION ALGORITHM

A simulation algorithm had to be developed which would check whether the proposed battery system is applicable to a real-life scenario. The algorithm checks the feasibility of such a system and tracks the energy level of the battery during a journey. The platform which was chosen for the algorithm is Matlab.

The main idea during the development of this software was that a typical journey of the containership should be split up between three ports. The transition from one port to another includes three different operations with different power demand profiles. During the port stay an assumption has been made that there is a shore supply (cold ironing) capable of providing the required power to charge up (fully) the battery during berthing. A simple schematic representation of this can been seen below, on Figure. 6.

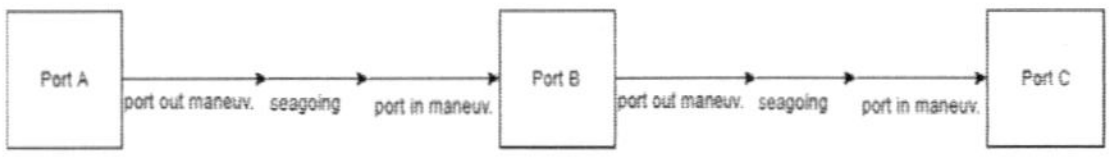

Figure. 6. Algorithm's schematic diagram

5. RESULTS - DISCUSSION

A way of testing whether the proposed configuration could be applied to a real ship had to be identified. This was done by utilising the voyage summary of the employed containership. The routes that were selected as case studies were very different among them in terms of manoeuvring and seagoing times. Some routes involved sailing in ECA zones where the use of MGO is required.

The results were a product of combining the simulation algorithm and the emissions model. The main idea when the simulation was executed, was that the number of generators running should be as low as possible and that their load factor should be as near as possible to the optimum point (around 85% of Maximum Continuous Rating). Also, when conditions were favourable, the battery was used as the only auxiliary power source in various stages of the journey, including sea passage. In this way, the ship would not be just a battery carrier but also a battery user similarly to a hybrid system.

In the beginning, the simulation was used to determine if a specific trip is feasible with the proposed battery system (no shortage of energy during all operations of the vessel). If it was feasible, then the optimum number of generators was found and at the same time the battery usage time was maximised not only on manoeuvring but also on sea passage operation. This was done under 2 scenarios (ship carrying only normal containers and ship carrying normal and refrigerated containers on-board) to account for the different power demand profiles. After the power demand conditions with the proposed system were satisfied, the emissions model was employed to calculate emissions with the parameters obtained from the simulation algorithm. One of these parameters was also the time of sole operation of the battery system, during all operations of the vessel, where no emissions will be produced.
To have a good comparison, emissions with the existing power configuration of the ship were calculated with inputs the number of generators running and their associated load factor during the various operations of the vessel. The emissions, with and without the proposed configuration for both normal containers and maximum refrigerated containers loading scenarios, were

calculated. There is a significant emissions reduction with the proposed arrangement. It should also be noted that emissions were calculated based on a scenario of a fully loaded vessel.

Table 2. Emissions reduction for a case study trip

Melbourne – Tauranga 1692.5 nm						
	Normal Case			Reefer Case		
	Business as usual	Battery	Reductions %	Business as usual	Battery	Reductions %
CO_2 (t)	81.47	64.84	20.41	697.42	681.8	2.24
SO_2 (t)	1.09	1.04	4.67	10.73	10.71	0.21
NOx (t)	1.43	1.12	22.06	12.06	11.73	2.72
Battery Usage @ seagoing (h)	-	18.8		-	3.07	

Emissions during all stages of the trip for both normal and reefer container loading conditions will be presented below. The analysis will commence for the normal case and the reefer case will follow. The analysis is synchronised with the chronological order of the different operations.

Table 3 illustrates the emissions for the manoeuvring out operation in the port of Melbourne. The emissions with and without the proposed modification are shown. As it is visible emissions during this stage of the trip are zero when using the proposed battery arrangement because the manoeuvring is 6 hours and, as it can be recalled from *Chapter 2.1* the battery system was designed to be able to provide energy for 8 hours of manoeuvring when the ship carries only normal containers and no refrigerated ones. Hence, the battery can provide all the energy needed by itself without any generators running.

Table 3. Normal manoeuvring out - Detailed emissions

Normal Manoeuvring out		
	Business as Usual	Proposed Modification
CO_2 (t)	6.5795	0
SO_2 (t)	0.0041	0
NOx (t)	0.1127	0

During the sea passage of the ship, the battery is charged up and then works as the only auxiliary power source for almost 19 hours. The emissions reduction between the two arrangements is small in this case. It can be explained by the fact that the one working generator on seagoing operates at a higher load factor (when compared with the current configuration) to provide energy to charge the battery after the ship has left Melbourne and before its arrival to Tauranga. This increase in the load factor indicates increased emissions, but battery usage (indicating no emissions) on seagoing compensates for these increased emissions. As a result, overall emissions are less. This is visible in the table below:

Table 4. Normal seagoing - Detailed emissions

Normal seagoing		
	Business as Usual	Proposed Modification
CO_2 (t)	67.47	64.84
SO_2 (t)	1.0833	1.041
NOx (t)	1.19	1.12

During the manoeuvring in on Tauranga, which is 6.1 hours, generators are turned off and the battery is the sole energy provider. Therefore, emissions are again zero for this operation of the vessel.

Table 5. Normal manoeuvring in - Detailed emissions

Normal manoeuvring in		
	Business as Usual	Proposed Modification
CO_2 (t)	7.42	0
SO_2 (t)	0.0046	0
NOx (t)	0.127	0

When refrigerated containers are on-board, the battery is not able to provide the whole energy required on manoeuvring out operation. Therefore, two generators must assist the battery in meeting the power demands of the vessel. Nevertheless, two less generators will be operating compared to the BAU (business as usual) scenario, having as a result reduced emissions. The emissions for the first stage of the journey are illustrated below:

Table 6. Reefer manoeuvring in - Detailed emissions

Reefer Manoeuvring out		
	Business as Usual	Proposed Modification
CO_2 (t)	14.26	7.3
SO_2 (t)	0.0089	0.0045
NOx (t)	0.2399	0.122

Table 7. Reefer seagoing - Detailed emissions

Reefer Seagoing		
	Business as Usual	Proposed Modification
CO_2 (t)	667.0792	666.27
SO_2 (t)	10.7096	10.69
NOx (t)	11.5588	11.48

During the sea passage, which is 130 hours, the battery operates for 3 hours as the sole energy provider for the auxiliary energy demands of the vessel. In addition, during the operation of the auxiliary engines one less diesel generator will be operating compared to the number of generators operating with the current power configuration. Table 7 presents the emissions. The reduction for this stage of the trip is negligible.

Similarly, to the manoeuvring out, emissions for manoeuvring in are less than those with the current power configuration because two less generators will be operating. These reduced emissions are almost the same to the manoeuvring out since the duration was 6.8 hours.

Table 8. Reefer manoeuvring in - Detailed emissions

Reefer manoeuvring in		
	Business as Usual	**Proposed Modification**
CO_2 (t)	16.0826	8.23
SO_2 (t)	0.01	0.005
NO_x (t)	0.2705	0.1376

6. LIFE-CYCLE ANALYSIS

As the final part of this design project, the potential profitability of removing one diesel generator and replacing it with a containerised battery system, was investigated. The methodology that was followed consisted of three main stages: 1) Identification of the costs and the savings associated with the proposed configuration, 2) Research on the prices of the individual components of the battery system, 3) Calculation of the Net Present Value (difference between the present cash inflows and the present value of the cash outflows). However, it should be noted that the below analysis is an approximate method of determining the potential profitability of the proposed design, since a variety of assumptions were used in the calculation process due to lack of available data.

6.1 COSTS AND SAVINGS IDENTIFICATION

After investigating all the practical costs and savings associated with the removal of one D/G and the fitting of the containerised battery system, the following five and seven, saving and cost parameters respectively, were identified (Table 1).

6.2 PRICE RESEARCH

The money gain (savings) that would result from the removal of 1 D/G was broken down into the following parameters: 1) the price of a 1600 kW D/G, 2) the annual lubricating oil cost, 3) the annual repair and maintenance cost of the D/G as well as 4) the money saved by burning less fuel due to the use of battery power throughout the manoeuvring operation. In respect of the price of a 1600 kW D/G, an indicative price would be approximately 500,000 USD. Regarding the annual lubricating oil cost an approximate value of 10,000 USD was calculated, assuming 10 tonnes of daily lubricating oil consumption.

Additionally, a price of 20,000 USD was considered for the annual repair and maintenance costs, since this value represents an indicative price of a complete overhauling of one D/G regarding the spares needed to be used once no serious damages are present. Finally, the fuel cost saved in ECA and non ECA areas would be variable and would depend on the case study trip investigated.

The money that would need to be invested in order the D/G to be replaced with a containerised battery system, mainly would consist of the cost of the battery packs themselves and the cost of the various components required for the battery system. Since, there were not available prices for the components and the battery packs, a worst case scenario price of 10,000,000 USD was assumed for the battery packs and approximate values for the other components were also assumed as it was illustrated in Table 9. Finally, the cost of charging the battery system from an external on-shore source (cold ironing) cost was calculated, assuming the cost of electricity being 0.07 USD/kWh and that the vessel carries out 70 trips annually.

Table 9. Identified cost and saving parameters

Project Cost Analysis			
MONEY GAIN DUE TO THE REMOVAL OF 1 D/G			
1) Price of 1, 1600 kW D/G:	500,000	USD ($)	(ship's life)
2) Lub. oil cost of 1 D/G:	10,000	USD ($)	(per year)
3) Repair + Maintenance cost (D/G):	20,000	USD ($)	(per year)
4) Fuel cost saved (MGO):	Variable	USD ($)	(per trip)
5) Fuel cost saved (HFO):	Variable	USD ($)	(per trip)
TOTAL MONEY SAVED:	Variable	USD ($)	
MONEY LOSS DUE TO THE ADDITION OF THE BATTERY SYSTEM			
1) Price of the Battery Packs:	10,000,000	USD ($)	(ship's life)
2) Price of Battery Management System :	20,000	USD ($)	(ship's life)
3) Repair + Maintenance cost (Battery System):	20,000	USD ($)	(per year)
4) Cost of Cabling:	40,000	USD ($)	(ship's life)
5) Cost of Transformers:	316,501.6	USD ($)	(ship's life)
6) Cost of Inverter/Rectifier units:	50,000	USD ($)	(ship's life)
7) Cost for electricity (charging from port external source):	45,717	USD ($)	(per charge)
TOTAL INVESTMENT:	Variable	USD ($)	

6.3 NET PRESENT VALUE RESULTS

In conclusion, a negative Net Present Value, which was found to be −7,964,999.8 $, indicates that the projected earnings generated by the proposed design will not exceed the anticipated costs, suggesting a financial loss. Nevertheless, it should also be stated that if the battery price assumed (10,000,000 USD) is reduced to 1/5 then the Net Present Value will become positive, transforming the proposed design into a profitable investment. In general, a wider use of containerised battery systems will be observed in the maritime industry if several improvements in the cost-performance and reductions in the price of the battery systems, are achieved. In fact, as it is stated in a recent study [10], the cost of battery systems is expected to decrease in terms of $/kWh, which will impact the economic viability of such investments. However, at the moment the motives for introducing such a system are purely environmental.

7. CONCLUSIONS

This paper has demonstrated that emissions reduction is possible through the installation of a battery system on-board an ocean-going vessel (containership). A battery system has been designed, capable of providing 10.30MWh of useful energy, when fully charged. This energy corresponds to the auxiliary demands for 8 hours of manoeuvring for the employed vessel (*"IRENES REMEDY"*, IMO: 9315850) when the ship carries only normal containers and no refrigerated ones. For a scenario of a manoeuvring operation being more than 8 hours, a diesel generator needs to be switched on to assist in meeting the auxiliary energy demands of the vessel.

In addition, it has been decided the battery system to be containerised in order not to alter the current arrangement of the vessel as well as for practical purposes (loading and unloading the battery system using gantry cranes). The 20ft refrigerated container has been selected as the type of container in order to control the environment (temperature & humidity) within the battery space of the container. Based on calculations performed 20 such containers would be required for the whole battery system. Within each battery container there would be a transformer (step-up & step-down) as well as an inverter/rectifier unit which will facilitate the connection of the battery system with vessel's switchboard.

Once all the technical characteristics of the battery system have been investigated, "ANKO" load planner software was utilised to ensure the stability and the structural integrity of the vessel after loading the battery system. It has been found that stability values as well as shear forces and bending moments, acting on the ship, are all within the limits indicating that the battery system would not negatively affect the wellbeing of the vessel, under a typical loading scenario.

Moreover, through the use of the reefer optimisation model it was obtained that once the battery system would be adopted 53 TEU and 310 FEU refrigerated containers can be accommodated on-board. By having these refrigerated containers on-board it was ensured that 3 generators can successfully meet the energy demands (as well as charge the battery) on seagoing and during manoeuvring the energy demands will be met by combining the energy from the battery system and from 2 running generators (for 8 hours of manoeuvring operation). An algorithm has also been developed on MatLab which tracks the energy level of the battery during the various operations of the vessel (manoeuvring out, seagoing, manoeuvring in) and checks whether the auxiliary energy demands of the vessel are met during these operations. Utilising this algorithm, it was deduced that when the battery system has drained, 64 seagoing hours are required to fully charge it, when the vessel carries only normal containers, whereas when the ship carries normal and refrigerated containers 97 seagoing hours would be required. If seagoing times are higher than the ones stated above, then the battery system can also be used on seagoing in order to cover, on its own, the auxiliary energy demands of the vessel. On the other side, if seagoing times are less than the ones stated above, then the battery system would not be fully charged after seagoing operation. However, that might not be a problem because the next manoeuvring operation might not require a fully charged battery.

Utilising vessel's voyage summary and by investigating 20 case study trips, it was calculated that the average emissions reduction with the battery system on-board would be 19 %, when the ship would carry only normal containers and 4 % when the ship would carry normal and refrigerated containers.

Finally, through the financial analysis conducted it was deduced that such an investment would result to a loss, primarily due to the high price of the battery system. Using a discount rate of 10% and an interest period of 20 years, NPV was calculated to be -7,964,999.8$. On the bright side though, the price of battery systems is expected to decrease making such investments economically viable. However, at the moment, the motives for adopting such an investment are purely environmental.

8. ACKNOWLEDGEMENTS

The above paper was completed and submitted by the group as part of the module MAR3033 Marine Engineering Design at Newcastle University. The authors would also like to thank Dr Kayvan Pazouki and Dr Rose Norman who were instrumental in getting this study completed by giving their valuable advice.

9. REFERENCES

1. IMO. (2008). Resolution MEPC.176(58) *Amendments to the annex of the protocol of 1997 to amend the international convention for the prevention of pollution from ships, 1973, as modified by the protocol of 1978 relating thereto.* [online] Available at: http://www.imo.org/blast/blastDataHelper.asp?data_id=23760&filename=176(58).pdf [Accessed 6 Dec. 2016].
2. IMO. (2016). International Maritime Organization. [online] Available at: http://www.imo.org/en/Pages/Default.aspx [Accessed 16 Oct. 2016].
3. DOO, H-W. and LEE Y-C. (2013) '*The Study On Ensuring Effectiveness of IMO Instrument Regarding GHG Emission from Ships - Focusing On MARPOL73/78 Annex VI*', Journal of Navigation and Port Research 37.5 (2013): 511-517. Web.
4. PARKER, S., RAUCCI, C., SMITH, T. and LAFFINEUR, L. (2015). '*Understanding the Energy Efficiency Operational Indicator: An empirical analysis of ships from the Royal*

Belgian Shipowners Association'. [online] UCL Energy Institute, Royal Belgian Shipowners' Association. Available at: http://www.lowcarbonshipping.co.uk/files/Ben_Howett/RBSA_final_main.pdf [Accessed 7 Apr. 2017].
5. KEVIN, L., GILLIAN, R., ANTHONY, W. (2006) '*Emission Inventory Guidebook Version: 3.4*'. London, UK: Lloyd's Register of Shipping.
6. ANONYMOUS (2017). Data provided by a shipping company "TCM".
7. MARINETRAFFIC.COM. (2017). IRENES REMEDY (MMSI: 636014740) Ship Photos | AIS Marine Traffic. [online] Available at: https://www.marinetraffic.com/en/photos/of/ships/shipid:754972/#forward [Accessed 7 Feb. 2017].
8. CORVUSENERGY.COM. (2017). Corvus Energy Storage System ESS:. [online] Available at: http://corvusenergy.com [Accessed 20 Oct. 2016].
9. ANONYMOUS (2017). Data provided by an engineering company "Helintec SA".
10. WU, P. AND BUCKNALL, R. (2016). *Marine Propulsion using battery power*. London: Department of Mechanical Engineering, University College London.

10. AUTHORS BIOGRAPHY

- Dimakopoulos Ioannis is a postgraduate student at Newcastle University.
- Gkinis Spyridon is a postgraduate student at Newcastle University.
- Lampris Nikolaos is a postgraduate student at Newcastle University.
- Menegakis Petros is a postgraduate student at Newcastle University.
- Panagoulias Georgios is a postgraduate student at Newcastle University.

WIND ASSISTED SHIP DESIGN ANALYSIS AND OPERATIONAL CONSTRAINTS

R Eggers, and **G Gaillarde,** Maritime Research Institute Netherlands (MARIN), The Netherlands

SUMMARY

Wind assisted ship propulsion is a subject of research at MARIN anticipating and supporting a wider uptake in the shipping industry. This paper discusses the possible transformation of the industry and the existing and upcoming research methods to support successful projects. Studies to predict fuel/emission savings are quite common by now and should be on a path of standardisation. A description is given for three different methodologies that can be used depending on the development stage of a new project. Whereas studies are common in this field, there is certainly a need to define what modelling needs to be done. Research into operational constraints when in transit with a wind assisted ship is hardly done so far. The authors propose a few relevant topics that should be dealt with.

1. INTRODUCTION

Climate change has drastically changed the societal pressure and political decisions on energy supply. Possible new scenarios can be written today, as wind of changes are blowing in the energy supply and transport. While each of us likes to believe that sustainable solutions are the only future, it is our common responsibility in the maritime industry, to make it happen. Upcoming technical challenges are great, promises are high and the worst scenario would be to disappoint the maritime sector with solutions not delivering on their expectations. It is then of highest importance to prepare sufficient research investments and cooperate at development and implementation level to make it happen. Lessons can be learnt from the current development of wind turbine energy generation, a high-growth industry which goes from land-based to sea at this moment. They have reached in a record time a decrease in costs to make it a viable industry.

MARIN, as a knowledge contributor for maritime research and innovation, is ready to assist in reaching the necessary milestones for sustainable sea transport. Part of the current developments and upcoming challenges are presented hereafter.

2. BACKGROUND AND AMBITIONS

2.1 WIND AS AN ENERGY SOURCE

Wind is intermittent and velocities constantly change. It often isn't available when needed most. Above all, it requires significant investment to transform into energy without a short term return. Those are the strongest arguments to abandon any development on use of wind as energy resource.

Despite those constraints, engineering solutions are being developed to capture such energy, distribute it where needed and likely make it profitable.

Wind is set to be the backbone of the future power system, meeting almost 30% of Europe's power demand by 2030, according to the International Energy Agency IEA. There is no doubt that there is energy in the air. The challenge remains to capture it.

Going from land-based to sea-based will allow to capture even more wind. Offshore Wind Energy is growing strongly in many countries. An earlier study from 2015 (Mast, Rawlinson & Sixtenssion, 2015) predicted significant developments, as shown in Figure 1. The expected capacity in 2030 was 65 GW.

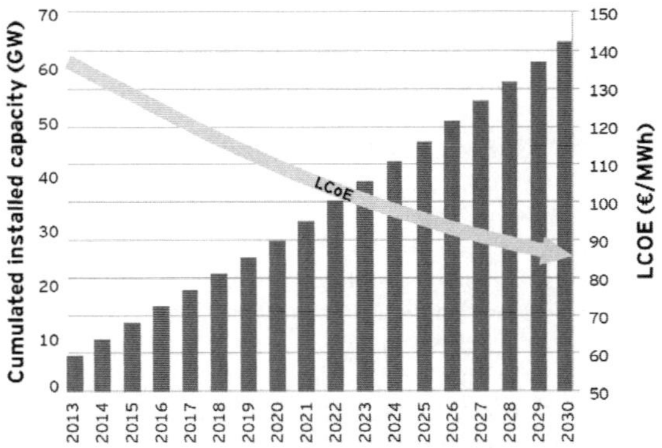

Figure 1: Cumulated Offshore Wind capacity

But things are changing fast, in particular thanks to the Paris agreement. It is expected today that offshore wind in the Netherlands could provide up to 50% of the total electric energy usage in 2030 (Tki Wind Op Zee, 2017).

At European level, offshore wind is expected to provide a total of 230 GigaWatt by the horizon 2030-2050, from which possibly 180 GW only in the North Sea. This is by far higher than the scenarios written in 2015.

2.2 WIND THRUST FOR SHPPING

Those remarks are far from the direct need in transport and shipping, but it shows that with a combined action of visionary governments, smart engineering and daring investors, a transition is possible in terms of energy supply. We see today concepts and products rising to transform wind into thrust onboard ships. Without doubting that the energy is available, the challenge is to develop designs and systems to capture such energy, without compromising the operations of the vessels.

A major contribution by wind as energy for sea transportation is also put forward by different studies, such as presented by SHELL in its strategic document: "2050: Fossil or sustainable" as shown in Figure 2. A study be CE Delft (Neilssen, et al, 2016) as commissioned by the European Commission also highlights the high potential of wind propulsion on ships (along with barriers that are to be overcome).

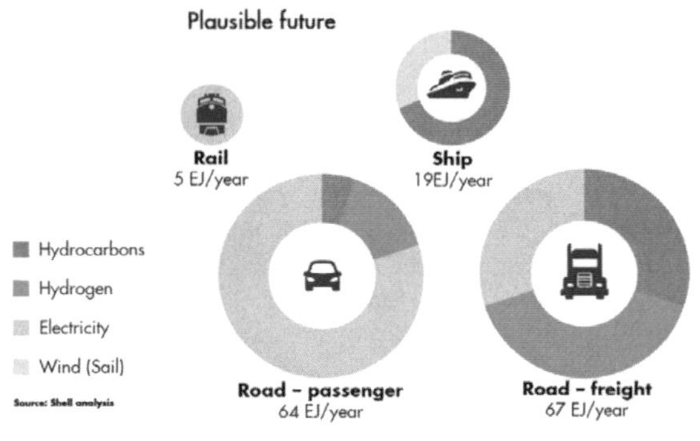

Figure 2: Possible energy supply per transport type *(source: SHELL analysis)*

Two approaches are going to prevail in the development of wind assistance:

- A refit market, where existing ship units will be equipped with standardised wings. They will probably not produce the most efficient solution in terms of wind propulsion but at least will be a low risk for the ship owners to "try" it. Lease options of portable systems or refit installations will appear.
- A new built market, where the ship design will include at concept level this mode of propulsion. At this moment, wind units are added to existing designs, but they will be much better integrated into the hull and superstructure when the ship is designed for it, also in terms of aerodynamics.

An additional development will certainly be on the supplier side. At this moment, one system is dominant on the market, but more research will likely introduce new systems and new players on the market.

3. METHODOLOGY TO EVALUATE WIND ENERGY PERFORMANCE IN SEA TRANSPORT

While the amount of prototypes to capture wind energy onboard ships is still limited, the feasibility studies to prove its potential have been numerous in the past years. All parties claim that today demonstrators must come to upscale those solutions.

No doubt about it, but we must also realise that demonstrators that will not meet the expectations will also jeopardise the development of this additional propulsion system. It is then of major importance to follow prescribed steps and make proper evaluation which is shared within the industry.

The steps that may be taken are herewith highlighted. These steps were followed in the INTERREG projects Wind Hybrid Coaster MariTIM (Maritim, 2016, Eggers, 2016) , SAIL (Schwarz-Röhr, 2015) and small research for the industry by MARIN. The example figures in this paper also come from these projects. It is further noted that MARIN is in the process to start a Joint Industry Project "WInd Ship Propulsion", "WiSP JIP" (MARIN), which, amongst others, intends to cover this topic.

3.1 WIND AVAILABILITY

The first stage to consider wind assisted ship propulsion may be at an owner owner/operator when no design is yet available, but only some intended routes or areas. Even without that design there are already relevant considerations. As wind assistance requires wind, a first analysis can be done on the wind speeds and angles over deck that may be expected. Hindcast weather databases can be sampled for the specific route. As shown in Figure 3 the results show the frequency of occurrence of wind angles and speeds. If the results show that the apparent wind on deck is only head wind, then it may be decided that the route is not suitable for wind assistance or the ship speed is too high. Assuming that this is not the case, the predominant wind angles and wind speeds already give a first indication which type of devices may be most suitable and which dimension may be appropriate. Some devices are better in stern quartering wind while other are more efficient in bow quartering wind.

3.2 STATISTICAL PERFORMANCE PREDICTION ON A GIVEN ROUTE

If the first stage was successful and some likely alternatives for the wind propulsion are available, then it is time for the second step: a statistical performance prediction. The same wind data as from stage 1 is used. However, at this point it is combined with the performance of the wind propulsion. The latter is obtained in a "PPP", Power Prediction Program (after the common acronym "VPP", Velocity Prediction program for sail boats. The prediction shows the power (or propeller thrust as in Figure 4) that is required to maintain a fixed ship speed for certain wind angles and velocities. By multiplying the required power with the wind probability for the same wind conditions, one arrives to the average (reduced) power required to maintain speed.

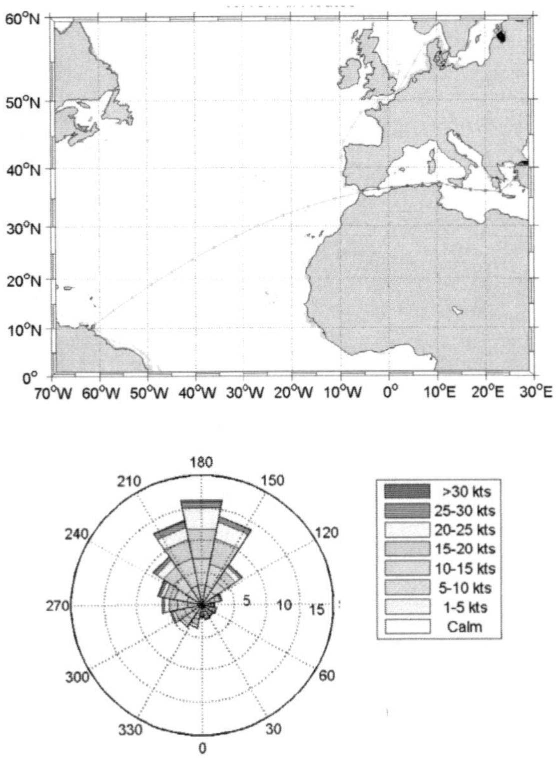

Figure 3: Evaluation of encountered wind conditions on a given route

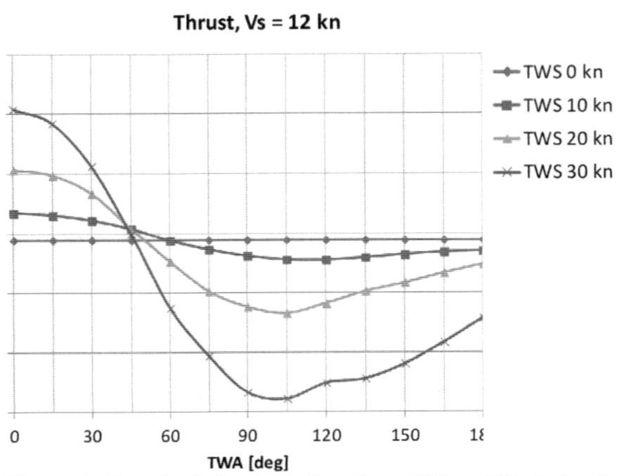

Figure 4: Required thrust as function of True Wind Angle (TWA) and True Wind Speed (TWS) for a fixed ship speed

EEDI implementation (IMO MEPC.1/Circ 815)

In its most simplest form this prediction could be based only on the wind propulsor. This is simple to obtain if wind coefficients are available (or generated) and only the settings (trim, rotor rotation rate, aspiration ...) is optimized. This is actually the way in which one can already demonstrate the savings in CO_2 emissions within the EEDI framework in the form of reduction of the EEDI. IMO provided the required calculation method in MEPC.1/Circ.815. An intermediate step in this process also yields the (reduced) fuel consumption.

Even though Circ. 815 does not discuss it in detail, it is noted that accounting only for the thrust generated by the wind propulsion is rather simplistic. At the IMO level it helps to keep things simple, however based on work mainly in the Wind Hybrid Coaster and SAIL projects it is MARIN's opinion that the method is too simple to give realistic estimates to make investment decisions.

Notably the method lacks in the following areas:

- Wind thrust larger than the required propeller thrust may lead to a negative (!) EEDI
- No constraints are mentioned in the use of wind propulsors (e.g. keeping heel within a certain angle)
- An increased resistance of the hull and rudder due to the leeway and rudder angle that are required to counter the side force of the wind propulsion
- Changed (decreased) efficiency of the entire remaining propulsion train at lower loads
- Further it is not mentioned whether the interaction between devices and the superstructure should be accounted for or in general what should be the quality of the used wind propulsion forces and required power to run rotors or suction wings

At least these aspects should be considered when preparing a prediction for wind propulsion. The exact impact and relevance of each item is discussed in some publications and presentations e.g. by Leslie-Miller (2017). However, this field certainly needs further research. One effort in this respect is the "WiSP JIP" as mentioned earlier.

Design variant and balancing aero and hydro forces

This is also the time to experiment with alternatives in wind propulsion, hull form and appendages. Aside from just performance, a related topic to be addressed is the steering balance. Most modern ships, when sailing at a leeway angle have the centre of effort of the side force somewhere near or even in front of the bow. The latter can occur due to an opposing force near the aft of the ship ("Munk Moment"). However, the centre of effort on the aerodynamic forces will be positioned somewhere within the length of the vessel, more likely closer to midship. So most probably there is a difference between these locations, which means that the rudder is required to obtain yaw balance. It must be ascertained that in order to keep balance the rudder will not need to go to excessive angles as that would decrease the overall steering ability. Furthermore, the distribution of side force over hull and rudder can be optimised (by positioning the wind propulsion and by using appendages on the hull). In short, the rudder is likely to be more efficient for small side forces as with it good aspect ratio and shape it is typically an efficient lift producer. However, at larger side forces, the rudder area simply becomes too small. Lift coefficients become too large and the rudder may be operating near stall. At that point the hull will be the more efficient side force producer.

3.5 FINAL VERIFICATION IN VOYAGE SIMULATIONS

Stage three, voyages simulations, would be the last step in the process. This step is intended to verify the performance with the best modelling available. Compared to typical predictions this step may be particularly important for wind assistance because demonstrating the performance in trials or actual service is more difficult than simply the speed/power performance in calm water. It will be hard in trials/service as it depends on some many independent variables and measurements always come with other disturbances (e.g. added wave resistance that will be inevitable with some wind).

The voyage simulations should also be conducted using hindcast data for wind. This time the waves and current are also added and simulations are done without reducing the data to statistics. At this point the data is used as available from weather models, i.e. dependent on time and location.

In order to reduce statistical uncertainty, simulations should be done over a relatively long duration (e.g. 5 years) and with many individual trips (e.g. one departure per week) being conducted within that duration. If this would not be done a few lucky (or unlucky) trips could dominate the results.

Modelling of propulsion train and energy flows

Voyage simulations are also a good tool to model and then check the usage of the propulsion train. As mentioned earlier, reduction of efficiency of the propulsion train at lower loads can be checked. E.g. this can be used to decide whether a diesel-direct or more advanced installation is required. Figure 5 shows the operational conditions encountered in a schematic engine envelope of a diesel direct installation with Controllable Pitch Propeller and Combinator. This particular example shows that loads can indeed be very low.

Sankey diagrams as in Figure 6 can be very useful to understand the relevance of the wind propulsion contribution and also the changes in efficiency. The example here is for a modest wind propulsion installation with Flettner rotors. Amongst other it shows here that the average leeway induced resistance for such installations can be modest when compared to the total (even if this resistance component may be very large at specific sailing conditions).

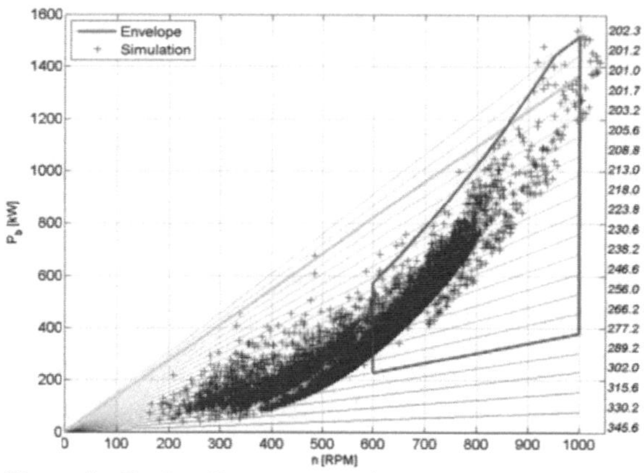

Figure 5: Engine diagram on a given route

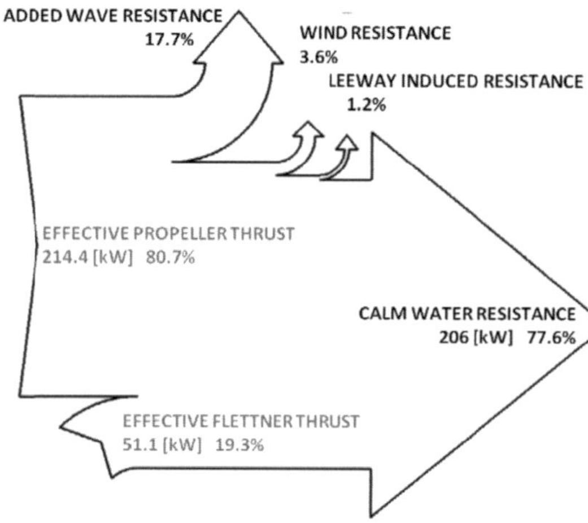

Figure 6: Average power balance at the level of thrust in service conditions

Route and speed changes and optimisation

Different than the statistical performance prediction, the voyage simulations can account for changes relative to a fixed route and speed which naturally occur due to voluntary speed loss/course changes (good seamanship and route optimisation) and involuntary speed loss (added resistance in waves). This includes sailing faster when the wind propulsion provides the majority of propulsion, which solves the problem that occurs for such in a condition in a PPP.

Thus, the merits of optimising the route and speeds for each individual trip can be demonstrated. Previous projects showed that the merits can vary greatly from route to route and from ship to ship. Previous projects at MARIN showed that re-routing was particularly interesting with variable winds from unfavourable directions for a ship with a large fraction of propulsion delivered by the wind, but not much in other cases.

Doing actual voyage simulations also allows to study the trade-off between average fuel/emissions savings and probability to arrive in time. Whereas it is perfectly possible to maintain an accurate schedule with wind assisted propulsion, the savings can be larger if some flexibility is allowed.

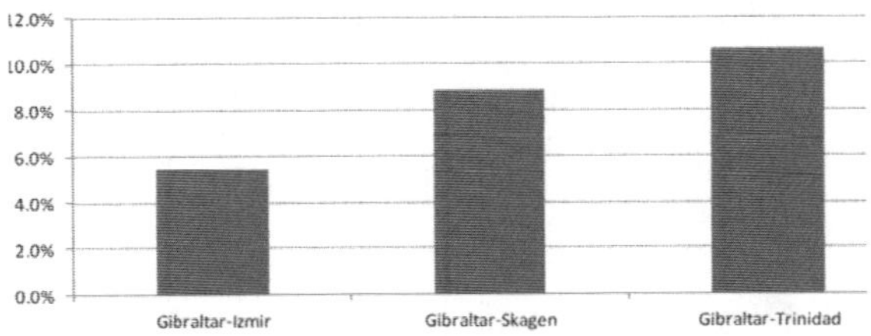

Figure 7: Fuel saving on several routes

4. CURRENT RESEARCH DEVELOPMENTS

As already mentioned previously, basic studies to determine performance for vessels with wind assisted propulsion have been numerous. However, there are many topics where further research is desired. Topics that will be addressed in the short term at MARIN are highlighted here.

4.1 DESIGN EXPLORATION

Whereas the performance of wind assisted vessels has been shown on a case by case basis, little systematic knowledge has been published about matching a wind propulsion device with a vessel (or vice versa). It is the combination of both that determines the performance. Optimisation algorithms could of course handle this as a black box. However, there are some specific tradeoffs that can be identified. Understanding these will greatly help the specification and design of ships with wind assisted ship propulsion. The following tradeoffs are under consideration:

- Wide versus deep hulls: At the same displacement a wide (shallow) hull will have a large roll stability, allowing it to be equipped with more wind propulsion capacity for the same maximum heel angle. At the same time a wide shallow hull will have a large leeway induced resistance and quite possibly also parasitic resistance. Thus, what is the "optimum" beam-to-draught ratio?
- Positioning wind propulsors (relatively) forward or aft: as mentioned earlier, the distribution of side force over hull and rudder determines the efficiency to generate side force and the steering balance.

As part of this research, efficiency of propellers in an oblique inflow condition requires revisiting past research by e.g. Gutsche (1975) with newer tools.

This research builds on a database of RANSE force calculations for a systematic series of hulls as conducted in a collaboration with Delft University of Technology, Dykstra Naval Architects and Damen Shipyards. Some of the parent hulls are shown in Figure 8.

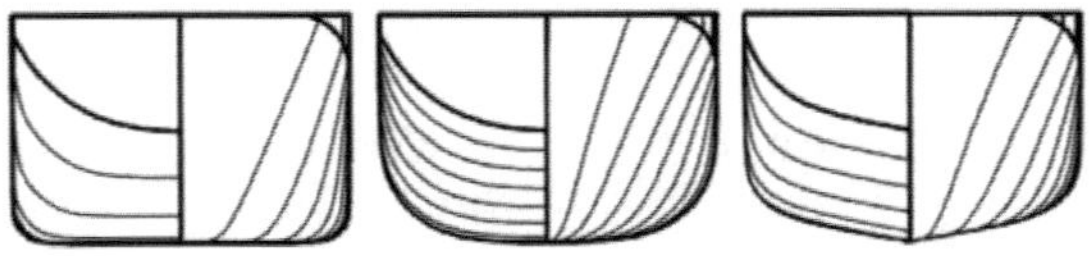

Figure 8: Example of hull form variation as already published by Van der Kolk *et al* (2017)

4.2 POSSIBLE OPERATIONAL ISSUES WHILE IN TRANSIT

A number of operational constraints for wind assisted ships while in transit have not yet received much consideration in general research and will also be part of further research at MARIN.

- The possibility to keep course in stern quartering wind and seas. Stern quartering seas, especially with wave lengths similar to the ship length and travelling at approximately the same velocity, are demanding for course keeping. When the wind is from the same direction then this is also very favourable for the wind propulsion, meaning that possibly there is hardly or no propeller thrust required. This means that the flow speed over the rudder(s) will be relatively low at a time when the highest rudder forces are required. Whether and when this really presents a problem and if it should be accounted for also in the speed-power prediction is to be investigated.
- Related to the previous item is the possibility to make manoeuvres, including avoidance of other ships. This relates directly to IMO resolution MSC 137(76) which a minimum performance in zig-zag and turning circle manoeuvres. The resolution does not state whether the performance needs to be met also with active wind propulsion. However, DNV-GL now has a draft procedure for certification of Flettner Rotors onboard vessels. There they require that the IMO rules for manoeuvring should also be complied with by ships with active wind propulsion. Thus, this should then be accounted for in the design stage.
- Finally, wind assistance will induce a heel angle if it cannot be compensated (e.g. by ballast water or roll stabilisation fins). The first question is: what is an acceptable heel angle? This will likely be different per ship type and may relate to cargo, crew and passengers. At the moment it is noted that various publications use no threshold at all while others do pick a number. If a threshold is used it tends to be 10 degrees. However, none of the studies known to the authors provided a firm basis for the selection of the

threshold value although for some ships it can really limit the achievable savings. Further, sails are known to increase roll damping. This means that aside from an increase of the (steady) heel angle, the (dynamic) roll angle due to waves may in fact decrease. An investigation of this issue and its impact on performance is to be done.

4.3 RESEARCH METHODS FOR OPERATIONAL ISSUES

The operational issues as described in the previous section may be assessed using time domain simulations. However, the required input is very detailed as it involves a lot of physics (restoring forces, damping, added mass effects) and control and this all needs to be covered in empirical models or databases. MARIN intends to not only do simulations, but to conduct "validation" scale model experiments which can then be used to compare against the simulations. The comparison will show what the way forward should be.

An example of the experimental set-up that could be used for wind assisted vessels was already applied for a few projects. Such a set-up was developed within the background research program of MARIN and was published for the first time in 2007 by Gaillarde and De Ridder *et al* (2007). Recently an improved version of this set-up was used for the AkzoNobel team in the 2017-2018 Volvo Ocean Race. In this effort a scale model of the yacht was equipped with a rig to allow a connection in the aerodynamic centre of effort. Lines were connected between this point and two individual actively controlled winches. A third winch was connected with a point on the bow to apply a small viscous model scale correction.

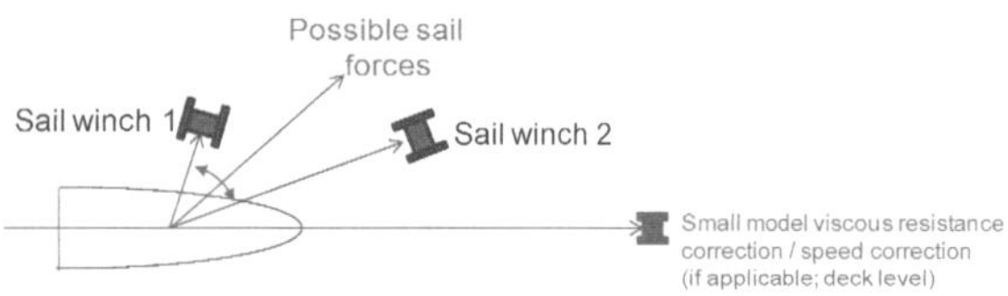

Figure 9: Top view of winch system to apply wind loads

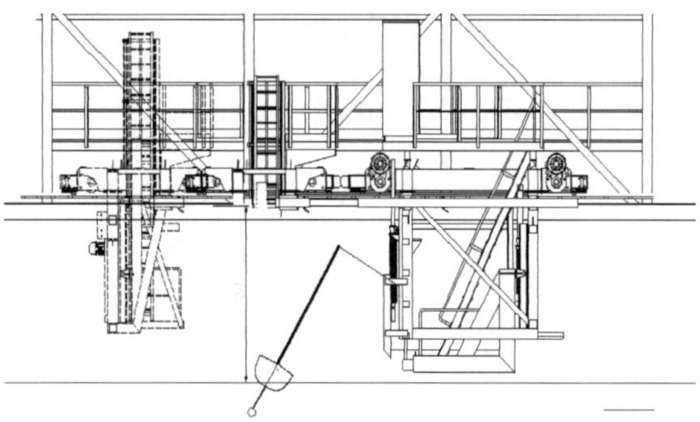

Figure 10: Front view of scale model connected to the winches on the carriage

The sail force winches were controlled to apply a continuously updated wind force. Using two winches allowed to vary the direction of the force vector between the two winches. For each test a true wind speed and wind angle relative to the longitudinal axis in the basin was defined. At each time instant, the apparent wind speed and angle were determined while accounting for the heading, leeway and speed. Based on the apparent wind angle and speed and the yacht heel and chosen sail trim settings, the force was looked up in the wind force database and applied by the winches. In addition an estimate of the aerodynamic roll damping was applied. This process was ran at each measurement time step. As far as known to the authors this set-up is unique in the world.

Figure 11: Picture of the VO65 model free running in bow quartering waves

7. CONCLUSIONS

As summarised from prior research and publications it is judged that using wind energy for ship propulsion has high potential, but both background research and applied research in projects is necessary for a bigger and successful uptake in the shipbuilding and shipping industry.

Based on MARIN experience in past projects, predictions on the performance of wind assistance ship propulsion can be done in three stages with increasing complexity and completeness:

- Stage 1: Wind availability on a route, looking at both the apparent wind angle and speed over deck as function of sailing speed
- Stage 2: A Power Performance Program (for a fixed speed) multiplied with the wind probabilities on a route. This is similar to the approach in EEDI MEPC.1/Circ.815, although the requirements in that approach are very flexible.
- Stage 3: Time accurate voyage simulations on a route. The intention in this last step is to deliver the best prediction possible, also accounting for

voluntary (re-routing and seamanship) and involuntary speed and course changes.

Whereas case studies have followed this path before, there are still open points with regard to the modelling. Aside from pure speed-power performance, the open topics for research also concern possible operational constraints in transit.

8. REFERENCES

1. MAST, E., RAWLISON, R., SIXTENSSION, C., '*Market Study floating wind in the Netherlands*', TKI Wind Op Zee, 6 November 2015
2. TKI WIND OP ZEE (Dutch "Topsector" research Wind at Sea), 2017 https://topsectorenergie.nl/en/tki-wind-op-zee ,
3. NELISSEN, D., TRAUT, M., KOEHLER, J., MAO, W., FABER, J., and AHDOUR, S., '*Study on the analysis of market potentials and market barriers for wind propulsion technologies for ships*', CE Delft, Delft, 16.7G92.114, November 2016
4. MARITIM (2016), Maritime Technologien und Innovationen, http://maritim-de-nl.eu/
5. EGGERS, R., '*Operational Performance of Wind Assisted Ships*', in 10th Symposium on High-Performance Marine Vehicles (HIPER), Cortona, Italy, 2016
6. SCHWARZ-RÖHR, B. *et al.*, '*Roadmap for Sail Transport: Engineering*', SAIL, SAIL Workpackage 3 Interreg IVB SAIL, October 2015
7. MARIN, '*Wind-Assisted Ship Propulsion Initial proposal for Joint Industry Project*' (http://www.marin.nl/web/JIPs-Networks/JIPs-public/WISP.htm)
8. LESLIE-MILLER, M., '*Quantifying the Attainable Fuel Savings with Wind Assisted Propulsion*', BlueWeek Spring Session www.blueforum.org , Den Helder, 2017
9. GUTSCHE, F., '*The Study of Ships' Propellers in Oblique Flow*', 1975
10. VAN DER KOLK, N. J., KEUNING, J.A., and HUIJSMANS, R.H.M. "*Hydrodynamics of Wind-Assisted Ship Propulsion Validation of RANS-CFD Methodology*" INNOVSAIL, Lorient, 2017
11. GAILLARDE, G., DE RIDDER, E.J., VAN WALREE, F., and KONING, J., '*Increasing Performance of IACC Sailing yachts through Seakeeping Study*', the 18th Chesapeake Sailing Yacht Symposium, Annapolis, March 2007

9. AUTHORS BIOGRAPHY

Rogier Eggers holds the position of Senior Project Manager and knowledge co-ordinator on Ships Manoeuvring at the Maritime Research Institute Netherlands. He started his career in 2008 at MARIN focusing on ship manoeuvring. Over the years he also covered seakeeping aspects. Finally, since about 5 years he is the main leader of background and applied research on wind assisted vessels at MARIN.

Guilhem Gaillarde is currently head of the Ships department at the Maritime Research Institute Netherlands. After graduating in marine engineering and naval architecture, he started his hydrodynamic career in 1998 at MARIN as Seakeeping project manager and evolved in multidisciplinary projects covering also powering and manoeuvring fields. He initiated the Natural Propulsion seminar in 2012 and is actually chairman of the Blue Forum which aims at initiating projects and discussing technical progress for sustainable marine activities and innovation. He is also a member of the RINA and of the SNAME.

LNG POWERED DRY BULK CARRIER

A Trakakis, Arista Shipping, Greece, **M Ioannou**, **M Penfold** and **E Kariambas**, American Bureau of Shipping, Greece and UK, **D Antonopoulos**, Wärtsilä, Greece

SUMMARY

LNG is one of the potential alternative fuel solutions to meet the specified IMO fuel sulphur content limits since it contains negligible sulphur. Furthermore, gas and DF engines utilizing the Otto combustion process can produce NOx emissions below the IMO Tier III limits. This paper discusses the challenges involved and how these were met, when considering LNG as fuel in an ocean-going dry bulk carrier, with focus on the main engine technologies and machinery, and their effect on the vessel design. The study was conducted by Arista Shipping, the American Bureau of Shipping (ABS) and Wärtsilä, and was based on a representative bulk carrier design, an 82,000 dwt Kamsarmax vessel, which was selected as the project platform due to its proven design and high cargo capacity.

NOMENCLATURE

2S	2-stroke (engine)
4S	4-stroke (engine)
BoG	Boil off Gas
CAPEX	Capital Expenditure
CFD	Computational Fluid Dynamics
CPP	Controllable Pitch Propeller
dwt/DWT	Deadweight tonnage
DF	Dual Fuel
ECA	Emission Control Area
EEDI	Energy Efficiency Deign Index
EGCS	Exhaust Gas Cleaning System
EU	European Union
FGSS	Fuel Gas Supply System
GHG	Green House Gases
HFO	Heavy Fuel Oil
HP	High Pressure
HSFO	High Sulphur Fuel Oil
IGC	International Code for the Construction and Equipment of Ships Carrying Liquefied Gases in Bulk
IGF	International Code of Safety for Ships Using Gases or Other Low-flashpoint Fuels
IMO	International Maritime Organisation
LNG	Liquefied Natural gas
LP	Low Pressure
LRM	Light Running Margin
LSFO	Low Sulphur Fuel Oil
mbarg	millibar gauge (unit of pressure)
mmBTU	Thousand thousand (i.e. million) British Thermal Units
MCR	Maximum Continuous Rating
MEP	Mean Effective Pressure
MEPC	Marine Environment Protection Committee
MGO	Marine Gas Oil
MMBTU	Million British Thermal Units
MN	Methane Number
MT	Metric Ton
NOx	Nitrogen Oxides
OPEX	Operating Expense
PTO	Power Take Off
SOLAS	Safety of Life at Sea
SOx	Sulphur Oxides
THC	Total Hydrocarbons
USD	United States Dollars

1. INTRODUCTION

In recent years, marine engine exhaust emissions legislation has become increasingly stringent. The upcoming global fuel sulfur content limit of 0.5%, effective from 1 January 2020, together with the existing ECA limits for fuel sulfur content of 0.1% and the Tier III NOx limits, impose significant challenges to the shipping industry. The options available to achieve compliance critically involve the selected engine technology and type(s) of fuel, two choices which then determine whether exhaust emission abatement technologies are necessary.

LNG contains negligible sulphur and it therefore meets all IMO regulations with regard to SOx emissions, as controlled by the fuel sulphur content limits. Furthermore, when combusted in gas or DF engines operating according to the Otto combustion cycle NOx emissions below the IMO Tier III limits are possible. On a calorific value basis, LNG prices are lower than traditional marine fuel prices. However, there are still challenges for the application of LNG as a fuel, notably fuel containment, fuel preparation systems, additional CAPEX and the LNG bunkering infrastructure. Given its strong potential, and despite the anticipated challenges, the use of LNG as a fuel is now extending beyond the application to LNG carriers, to other ship types such as bulk carriers and tankers.

This paper discusses the challenges involved in considering LNG as fuel in an ocean-going dry bulk carrier employed on varied international shipping routes and focuses on the main engine technologies and machinery space arrangements. The study is based on a bulk carrier design, an 82,000 dwt Kamsarmax vessel.

The study concentrated on the selection criteria and available technologies for the main engine and the respective machinery space arrangements, whilst taking into account the applicable Classification requirements, relevant statutory safety regulations (IGF code) and the shipowners operational requirements to deliver a highly efficient competitive solution, with increased reliability and redundancy characteristics.

The findings show that despite the higher CAPEX incurred for the application of LNG as a fuel, the payback period could potentially be a few years and provide a step forward in delivering a competitive, environmentally friendly merchant vessel with favourable total cost of ownership.

The study took into account all major factors that affect the selection of the propulsion system and the vessel design. Such factors included consideration of the applicable emission legislation, the various types of LNG containment systems and the location on board, together with the impact (stresses/stability) on the vessel design, the machinery requirements and the type of main engine to be adopted. Further considerations were given to CAPEX and OPEX, as well as to fuel prices, the fuel supply infrastructure, and to the regulatory framework. The LNG fueled bulk carrier design discussed in this paper received an Approval in Principle from ABS on 22 August 2016.

2. IMO SAFETY REGULATION

The IMO adopted the *International Code of Safety for Ships Using Gases or other Low-flashpoint Fuels (IGF Code)* by Resolution MSC.391(95) in June 2015. At the same time Resolution MSC.392(95) was adopted implementing amendments to SOLAS to make the IGF Code mandatory, introducing a new 'Part G' for ships using low flashpoint fuels and applying an entry into force date of 1 January 2017. The basic philosophy of the Code is to provide mandatory provisions for the arrangement, installation, control and monitoring of machinery, equipment and systems using low-flashpoint fuel to minimize the risk to the ship, its crew and the environment.

The IGF Code provides detailed prescriptive safety requirements for ships using natural gas as fuel under Part A-1 of the Code. This prescriptive approach is based largely on the extensive experience gained with the carriage and burning of LNG as fuel in gas carriers covered by the IMO IGC Code over many decades. It also applies reduced risk assessment obligations for the use of natural gas as fuel.

The majority of the existing gas fuelled fleet experience (non gas carriers) has been gained with ships built to the preceding IMO Resolution MSC.285(86) *Interim Guidelines on Safety for Natural Gas-fuelled Engine Installations in Ships*, which was adopted 1 June 2009. However, the LNG powered dry bulk carrier detailed in this paper has been assessed on the basis of compliance with the IMO IGF Code.

3. IMO EMISSIONS REGULATION

There are two main drivers pushing the consideration of LNG as fuel for marine propulsion systems. The first relates to air pollution emissions legislation and the second to the availability and price uncertainty of conventional liquid fuels in the future. IMO Annex VI limits sulphur emissions and the sulphate fraction of particulate matter through the fuel sulphur content limits of Regulation 14 and NOx emissions through Regulation 13. Both regulations have global limits and also local limits regulated through Emission Control Areas (Figure 1). With the adoption of GHG emission controls through the energy efficiency requirements adopted by IMO MEPC 62 in July 2011 there is now also the GHG incentive for the selection of low carbon fuels such as LNG. Dual fuel and single gas fuel engines have the potential to reduce CO_2 emissions.

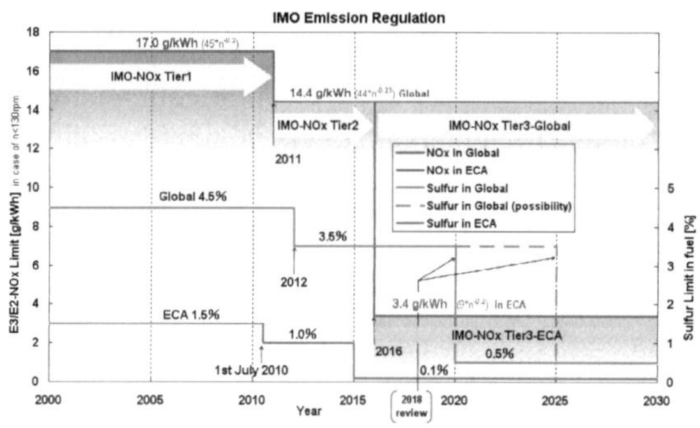

Figure 1. IMO Emissions Regulation (NOx and SOx)

Despite the recent low fuel oil prices, there remain long term question marks over future fuel prices, since a sudden shift in the demand of fuel type globally, from HSFO to 0.5% Sulphur fuel (Shell, 2017), estimated at a total of about 200 mil tonnes per annum, will have significant but not fully known consequences on the fuel bunker markets

On the other hand, the global abundance of gas, as epitomized by the unconventional North American shale gas supplies, provide the potential to deliver low cost LNG as a marine bunker fuel. This supply driver, along with the spectre of market based measures penalizing high GHG fuels, suggests that LNG can play a growing part in the marine fuel supply chain of the future.

3.1 ROUTES TO NOx COMPLIANCE

Shipowners, engine designers, builders or operators have three basic routes available to IMO Tier III NOx compliance:

- Conventional but advanced diesel engines running on either liquid or gas fuel using internal measures such as Exhaust Gas Recirculation (EGR);
- Conventional diesel engines running on liquid fuel using measures such as Selective Catalytic Reduction (SCR) after-treatment systems, which could be of the low pressure (LP SCR) or high pressure (HP SCR) type; or,
- Low pressure DF or gas engines operating according to the Otto combustion cycle that produce NOx emissions below the Tier III limits.

Each compliance option has its own advantages, disadvantages and CAPEX/OPEX impacts, which means the most suitable option is very much a ship and route/operating profile specific decision.

3.2 ROUTES TO SOx COMPLIANCE

The IMO fuel sulphur limit of 0.5% is applicable globally from 1 January 2020 and the 0.1% fuel limit in ECAs has been effective since 1 January 2015. Compliance with these SOx regulations is possible through two basic routes:

- The use of sulphur compliant fuels, which would include conventional liquid fuels such as MGO and other low sulphur fuels such as LNG; or,
- The use of Exhaust Gas Cleaning Systems (EGCS) on-board the vessel that reduce the exhaust sulphur emissions to an equivalent level as the fuel sulphur limits, when running on higher sulphur content fuels, typically using a wet exhaust scrubbing process.

The CAPEX and OPEX costs associated with EGCS may be prohibitive and installation can involve major vessel retrofit. Furthermore HFO with high sulphur content is not expected to be as widely available in the future as it is today. It is therefore questionable whether the number of vessels that will be fitted with EGCS by 2020 will be sufficiently large to form a solid market for the continued supply of HSFO.

4. MAIN ENGINE SELECTION

The performance of a vessel depends on many factors, one of the most important and complex being the choice of propulsion system to be installed on the ship. The propulsion system accounts for a large CAPEX percentage of the total shipbuilding cost, as well as a large portion of the annual operating costs in fuel consumption and maintenance. Therefore the selection of the most suitable engine is a crucial factor for shipowners and designers and requires thorough analysis so that the most appropriate propulsion system is selected at an early stage of the vessel design process. The analysis needs to take account of the vessels' design performance requirements, operating profile and recognise that the machinery selection has a major impact on the reliability of a vessel.

For the use of LNG as fuel, there are two distinct engine design concepts available from the major marine large engine manufacturers designed around the Diesel or Otto combustion processes. For the use of LNG the re-gassified LNG needs to be fed into the engine as gas at different pressures, depending on the combustion process. For the Diesel process this is high pressure (approximately 300-350 bar) and for the Otto process this is low pressure (approximately 5-12 bar). This by necessity means there are also significant differences between the Fuel Gas Supply Systems (FGSS) that are required for each engine concept. The different combustion options, run on different cylinder heat cycles and operate with very different characteristics.

The candidate LNG as fuel engines for this study included all possible options of main engines and main engine configurations available on the market, including 2-stroke high pressure and low pressure gas supply engines, and the possibility to adopt a 4-stroke, low gas supply pressure engine.

For the case of the 4-stroke option, the propulsion system selected consisted of two Wärtsilä 31DF engines providing power to a twin input single output gearbox driving a Controllable Pitch Propeller (CPP), delivering an MCR of 8,830 kW at 80.0 rpm. With this configuration, one or two engines can be configured to operate at an optimum efficiency point depending on the vessel speed. A variable speed mode of the engines can also be applied. Auxiliary electrical power (hotel load) can also be produced from these propulsion engines by the use of two shaft generators with frequency converters driven through Power Take Off (PTO) arrangements. With this configuration there is sufficient load capacity to cover the complete range of the vessels' propulsion and generating needs at any operating condition. Therefore the need for additional generator sets to be installed for auxiliary power production can be avoided.

An evaluation system was developed that captured the main trends of the systems involved. The main criteria that the engines were evaluated upon were environmental, performance, financial and operational. The data was collected from publically available sources and some of the findings are depicted in Figure 2.

From the evaluation matrix it was concluded that given the current status of the various technologies, the most suitable propulsion arrangement for the vessel under study was the proposed multi-engine arrangement with 4-stroke DF engines operating on low gas supply pressure and driving through a CPP gearbox. The key factors that led to the decision were the increased safety and redundancy from two engines driving the propeller, high efficiency (due to the high level of technology utilized in the Wärtsilä 31DF design), the clean and environmental friendly operation, and the well proven service record and reliable operation of similar DF engine types.

	2S HP DF	2S LP DF	4S LP DF
THC	😊	😐	😐
CO_2	😊	😊	😊
SOx	😊	😊	😊
NOx	☹	😊	😊
PM	😐	😊	😊
Power density	😊	😐	😊
Efficiency on Gas	😊	😊	😊
Efficiency on Diesel	😊	😐	😐
MN dependency	😊	😐	😐
Transient response	😐	😐	😊
Main engine cost	😐	😐	😊
Gearbox	-	-	☹
Cost of FGSS	☹	😊	😊
Fuel cost	😊	😊	😊
Maintenance cost	☹	😊	😐
Proven service record	😐	😐	😊
Safety / redundancy	😐	😐	😊
Propeller efficiency - LRM	😐	😊	😊
Manoeuvring	😐	😐	😊
Reserve elec. load carrying capacity	😐	😐	😊
Weight compared to standard vessel	😐	😐	😊

Figure 2. Criteria and evaluation scores for three DF propulsion systems

The layout of the propulsion system selected is shown in Figure 3 below.

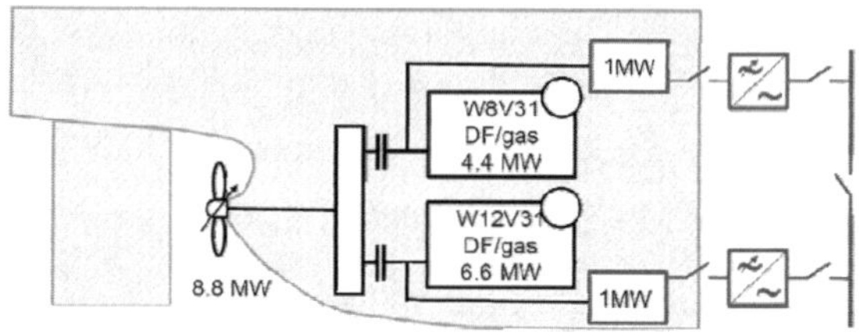

Figure 3. Schematic of propulsion system layout in engine room

The Wärtsilä 31DF engine (Figure 4) is a technology enabler for this selection because of the benefits from high efficiency due to the 2-stage turbocharging arrangement, and the NOx Tier III compliance when operating in gas mode which alleviates the need for complex emissions aftertreatment equipment. Furthermore the design of this engine has targeted reducing methane slip by addressing the following:

- Scavenging process, losses during valve overlap: engine has variable valve timing,
- Quenching effects in lean mixture zones: cold areas are avoided by employing high Mean Effective Pressures (MEP),
- Trapped methane in crevice areas: engine has optimized piston and cylinder design,
- Delayed combustion/misfiring: engine has advanced fuel injection technology.

Furthermore, due to the high efficiency the engine has a reduced GHG footprint, in particular in the range of 85% engine load the engine has about 25% less total CO2 emissions compared to the equivalent diesel version,. These characteristics make it the most suitable candidate for accomplishing the targets of the project.

Figure 4. Wärtsilä 31DF engine

5. VESSEL DESIGN

Due to the low density of LNG compared to conventional fuel oils, LNG requires much larger storage volumes for the same energy content. Furthermore, the long range operating profile of the vessel required a tank with a volume of 2,500 m^3 which would suffice for 40 sailing days at sea with full speed.

The position of the LNG tank on board could not be selected with the same degree of freedom as with conventional fuel, but instead, needed to comply with the provisions of the IGF code. As mentioned above, the IGF entered into force from 1 January 2017. The IGF Code includes alternative LNG tank location criteria using either deterministic or probabilistic methods; for this study the probabilistic approach based on the calculation of damage probability allowed a reasonably large tank to be accommodated on board within an efficient ship design.

There are several available technologies for the containment of LNG, all of them with very different pros and cons to consider. Major factors that influenced the LNG tank choice for this study were: volumetric efficiency, experience from existing installations, holding pressure and boil-off gas (BoG) rates/management, investment cost, cooling down practices, survey and inspection requirements.

A membrane type tank with holding pressure of 2.0 bar was found to be the optimum solution for the intended LNG volume. This design consisted of corrugated stainless steel with a thickness of 1.2mm which serves as the primary barrier being in contact with the LNG, welded on top of the insulation panel which integrates the secondary barrier between two layers of insulation. The membrane type allowed the maximum utilization of existing space between all containment systems, and at a lower cost and weight.

There are many ways to accommodate the LNG fuel containment system within the ship's hull, the initial approach being to increase the ship's length. This option offered several benefits, such as no loss of cargo space and no penalty on ship's draft, but does increase the ship DWT by 1,500 tonnes. This approach does however give the flexibility to adjust size depending on the particular needs of each owner, no negative impact on ship's stresses and stability, and could be applied for the conversion of young vessels should future conditions (e.g. price of 0.5% sulphur fuel) justify this move.

In many cases, bulk trades are not dependent on vessel draft, since draft restrictions apply at ports where low density cargoes are loaded and the quantity to be loaded is restricted by the total hold volume, and not by the DWT of the ship. Typically heavy cargoes, which require scantling draft, are loaded at ports where there are no draft restrictions. Therefore, the option to increase a ship's depth instead of length remains a possibility. Furthermore, considering the new Panama Canal dimensions, one may also opt to increase the ship's beam. We therefore see that each option requires different extra steel weight and the optimum solution depends on both the specific commercial requirements and the extra cost considerations.

Taking the aforementioned into consideration, the final design selected in this study is depicted in Figure 5, showing the location of the LNG tank of 2500 m^3 capacity and leading to an increase in vessels length of 10 meters. This solution allowed the proposed B.Delta82 LNG vessel to have a new DWT of 83.500t (including an additional 1,500t of cargo), with service speed of 14.5 kn at 7,100 kW.

The overall design details of the vessel include a length of 238m, a breadth of 32.26m, a depth of 20.3m, a draught design of 12.2m, a draught scantling of 14.45m and a DWT scantling of 82,000 tons.

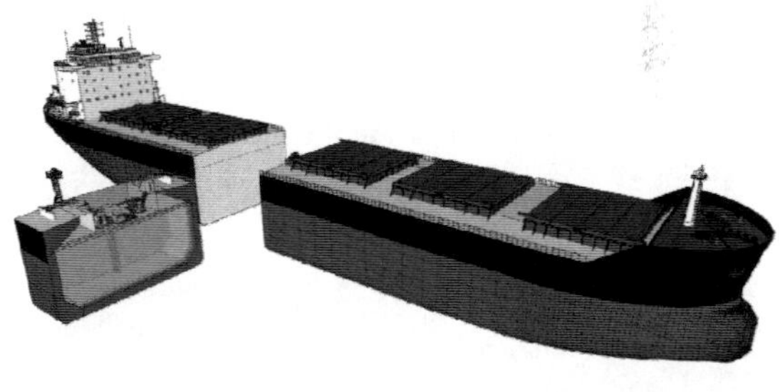

Figure 5. LNG tank on the B.Delta 82 bulk carrier [Gaztransport & Technigaz - GTT]

Due to the heat input to the tank, the liquid evaporates and builds up pressure (natural BoG). While at sea, the total gas consumption/requirement for propulsion and electrical loads is far higher than the natural boil off rate, therefore the BOG can be managed by the demand and the makeup gas is supplied from forced BoG. During stay at port, the natural BoG rate is slightly higher than the quantity required to cover the electrical loads alone. The expected boil off rate under conditions of high air and sea temperature, with the LNG tank being at maximum filling level has been estimated to be approximately 135 kg/h, whereas the electrical load consumption is estimated at approximately 100 kg/h.

Analysis has shown that for various LNG tank filling conditions (42%, 50%, 61%), which are typical for arrival conditions at ports where a long idle stay is expected such as South America, the gas pressure in the tank would not exceed 1.4 barg after a 40-day period in idle condition, i.e. load only for electrical power.

The current tank is able to hold pressure for more than 15 days without consumption and for more than 30 days with minimum hotel load.

It has also been determined that the LNG filling level at arrival will have a limited influence on the gas pressure increase. The gas pressure increases at a lower rate when the filling level is lower and indeed the absolute amount of BoG decreases with the tank surface area in contact with LNG (wet surface effect), which is dictated by the filling level. The combination of BoG consumption and fuel containment pressure capability therefore maintains LNG tank pressures within acceptable limits for the majority of operational scenarios.

In addition, ageing calculations undertaken during the study have shown that the quality of LNG will remain quite stable during idle periods, and that this phenomenon is not dependant on the LNG containment technology.

The last major influence on the vessel design, from the perspective of the selected propulsion system and the allowance for the LNG tank, involved the effect on the hull lines of the vessel. These were modified accordingly as shown in Figure 6. The specific twin-engine layout allowed for a more streamlined aft part, and further optimisation was achieved by positioning the engines above the gearbox centreline. Optimization and CFD calculations conducted in parallel by both ABS and the vessel designer Deltamarin, succeeded in recovering the extra losses incurred by the extra ship's length.

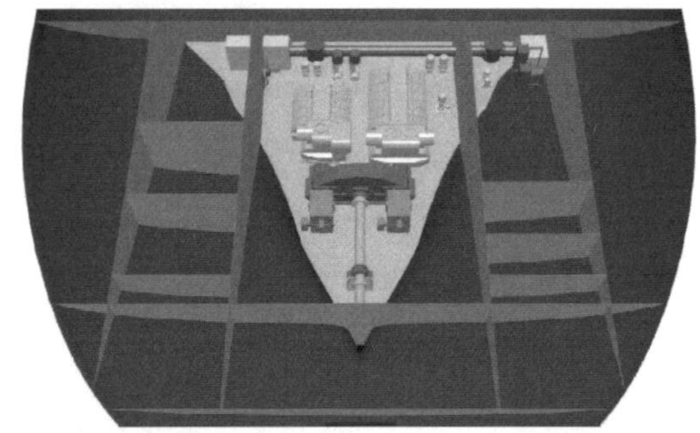

Figure 6. Aft part of the vessel accommodating the selected propulsion system [Deltamarin Oy]

6. CAPEX AND OPEX

The study also compared the machinery options on a CAPEX basis and included comparisons of a typical 2-stroke diesel engine running on liquid fuel and equipped with Tier III NOx technology (e.g. SCR), and the LNG fuelled alternatives. As previously indicated, the study considered two 4-stroke DF engines operating on the Otto combustion cycle in gas mode with low pressure gas injection, a 2-stroke DF engine operating on the Diesel combustion process in both oil and gas modes with high pressure injection, and a 2-stroke DF engine operating on the Otto combustion cycle in gas mode with low pressure injection.

The analysis shown in Figure 7 indicates that although the conventional vessel has the lowest initial cost, the proposed arrangement with two Wärtsilä 31DF engines provides the minimum extra CAPEX among all the considered options for the application of LNG as fuel with the aforementioned technical advantages.

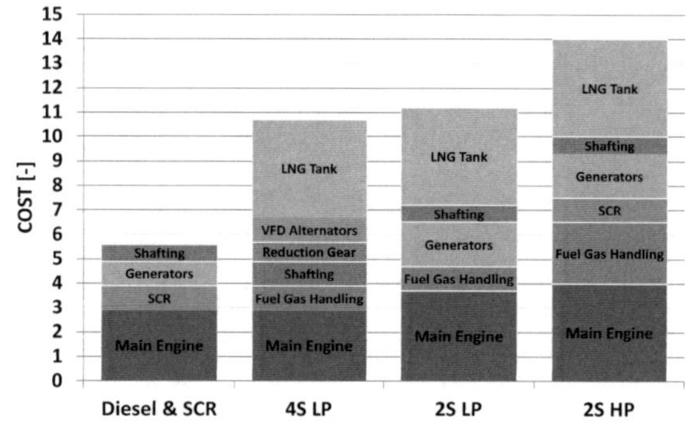

Figure 7. Dimensionless CAPEX comparison for Main Engine and all supporting equipment

6.1 FUEL PRICES

Fuel costs are the biggest factor affecting a ship operator's profitability. At the end of 2016, the purchase price of LNG was approximately 6% lower than that of HFO, on an energy content basis. While fuel and gas prices do fluctuate, the long term trend is expected to favour LNG as the least expensive choice. The fuel efficiency of modern gas-fuelled engines is also higher, with thermal efficiency ranging from 5 to 14% higher than conventional fuelled engines. The business case therefore needs to consider these factors and show that the difference in OPEX, predominantly through the cost of fuel, justifies the decision of adopting the LNG as fuel technologies.

Investigations were therefore carried out in order to determine whether reduced operating expenses (OPEX) could compensate for the increased CAPEX and justify the application of a 4-stroke or a 2-stroke DF engine. The maintenance costs of such engine installations is expected to be comparable to traditional solutions on conventional vessels. Piston overhaul is normally undertaken every five years but due to the cleaner combustion with gas the costs incurred could potentially be lower and the consumption of lubricating oil is also lower. Maintenance costs of the CPP and the reduction gear are not considered significant (for the 4-stroke application). Certain additional maintenance costs related to LNG have not been considered by the study, since similar costs would be accrued by vessels equipped with SCR and SOx EGCS technologies. Long term projections indicate that the price of LNG may range around 9 USD/MMBTU or even lower, including a 3.5 USD/MMBTU premium for delivery onboard (Waterborne Energy Inc, 2017).

Although it is difficult to predict the price of 0.5% sulphur fuel it is reasonable to assume that as the year 2020 approaches this will rise and become higher than current prices.
In order to examine whether the offset in CAPEX (Figure 7) could be paid back through the different fuel prices, an analysis was conducted comparing the LNG and LSFO prices across several price scenarios. Figure 8 shows that the payback period for the CAPEX could be achieved within 2-3 years, a scenario that clearly justifies the adoption of LNG as fuel technologies.

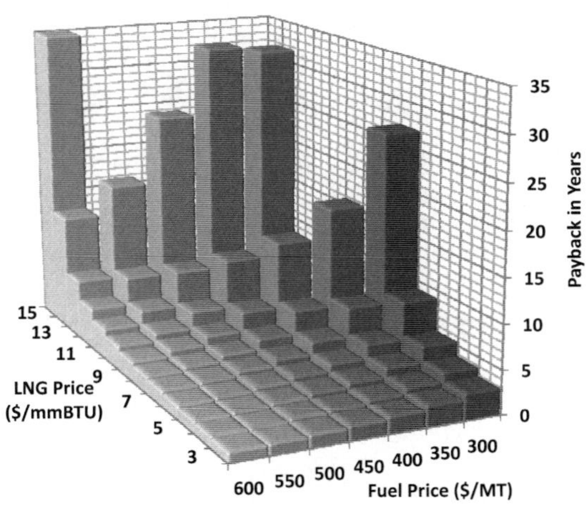

Figure 8. Payback period of additional CAPEX for different LNG/LSFO price scenarios

7. FUEL SUPPLY INFRASTRUCTURE

Conventional liquid fuels are readily available in almost all ports worldwide but for cost related reasons bunkering, in practise, takes place mainly at major ports, the trend is depicted in Figure 9.

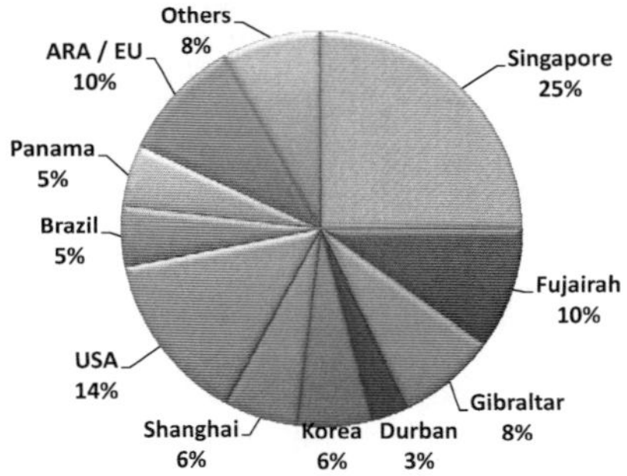

Figure 9. Bunkering uptake of vessels engaged in tramp service

The development of LNG bunkering supply networks in such ports is currently ongoing by major fuel supply companies with initial focus on areas such as the EU, Middle East, Singapore, and the United States.

It is therefore expected that LNG availability in such ports will soon be sufficient to accommodate the expected demand, and support the adoption of LNG as a bunker fuel.

8. CONCLUSIONS

The widespread adoption of LNG as a bunker fuel faces several challenges, including the investment required for the ships propulsion and fuel handling systems, the lack of bunkering facilities and infrastructure, uncertain LNG bunker prices and the necessary development of new international and local safety regulations, particularly with regard to bunkering and shore side LNG supply.

However, the benefits that it offers, for example, the significantly reduced EEDI attributable as a low carbon fuel, the compliance with SOx and NOx Tier III limits without any exhaust emission abatement technologies, and its abundant supply resources (proven reserves 6,900 Trillion cubic feet (US Energy Information Adminsitration, 2017)), can provide the stimulation to overcome these challenges.

The feasibility study described in this paper has demonstrated that through a novel approach to the machinery arrangements, adoption of 'state of the art' prime movers and novel pressurized membrane LNG tanks, it is indeed feasible to use LNG as a bunker fuel on an ocean-going dry bulk carrier. The study shows that the payback period for new construction is viable for the majority of the price scenarios considered. Despite the increased CAPEX compared to conventional ships, the benefits are numerous and demonstrate that the overall investment is expected to pay back in a very short time period with a favourable total cost of ownership.

Therefore, given the potential advantages of LNG as a bunker fuel, and the rate that the LNG infrastructure is developing, it is evident that there is a strong possibility that LNG will capture a significant percentage of the conventional fuel oil market in the near future.

9. REFERENCES

1. SHELL, www.shell.com/marine, April 2017
2. WATERBORNE ENERGY, INC, Aug 2017
3. UNITED STATES ENERGY INFORMATION ADMINISTRATION, www.eia.gov, 2016

10. AUTHORS' BIOGRAPHY

Antonis Trakakis holds the current position of Technical Manager at Arista Shipping. He is responsible for the technical management of a young fleet of bulk carriers. As Technical and Environmental Manager of Superfast Ferries he has successfully addressed all issues relevant to operation of ships in sensitive areas like the Baltic Sea.

He has worked with all kinds of internal combustion engines, ranging from gas turbines, to all sizes of 4-stroke, and 2-stroke engines, and one of his special concerns has been the formation and control of emissions, and engine efficiency.

Marios Ioannou is a Principal Engineer with the American Bureau of Shipping, in the Corporate Technology department. Prior to that, Dr. Ioannou has worked for many years for a number of engine manufacturers. In his latest role, with Winterthur Gas and Diesel (Wärtsilä) in Switzerland, he was an expert in the engine performance department and was involved in the development of the combustion systems of Diesel and Dual-Fuel Engines, with focus on both, the Diesel and Otto cycles. He was also the Performance Leader during shop tests and Type approval tests at licensee sites and sea trials in China, Korea and Japan.

Dionysios Antonopoulos is a Naval Architect & Marine engineer, currently holding the position of the General Manager, Marine Solutions Sales in Wärtsilä Greece. Mr. Antonopoulos worked for 5 years in Wärtsilä Switzerland, in the department of marketing and newbuilding projects and then appointed as Portfolio Manager for 2-stroke engines, still within Wärtsilä Switzerland, where he cooperated with major Korean and Chinese shipyards for aspects of vessel propulsion and engine development. Mr. Antonopoulos returned to Greece as Technical Manager for 2-stroke engines responsible for Greece and Cyprus and then appointed to the current holding position. Prior to Wärtsilä he worked for a shipping company and he also has seagoing experience.

HYBRID ENERGY - AND PROPULSION SYSTEM FOR VESSELS IN TIMETABLE OPERATION

M Einsiedler, Shiptec AG, Switzerland

SUMMARY

When inland vessels are either overhauled or designed from scratch, the close evaluation and specific optimization of their energy consumption from onboard systems as well as propulsion is increasingly important. On conventional ships, the rated power of every propulsion engine or generator must correspond to required peak power. For vessels that operate intensive, timetabled services and are only equipped with conventional diesel engine, this creates a degree of energy inefficiency due to suboptimal transient processes. The hybrid energy and propulsion system, which is described in this paper, uses smaller components in combination with energy storages. This makes it possible to break peak demand and more evenly distribute energy use/production. The system architecture is designed to maximize the energy efficiency of all components and improve the full integration of power producers and consumers. This enables a reduction of operation costs and maintenance expenses, while significantly lowering CO_2 emissions. Compared to similar ships running on the same routes, fuel savings should amount to 25-30%.

1. INTRODUCTION AND INITIAL POSITION

When inland vessels are either overhauled or designed from scratch, the close evaluation and specific optimization of their energy consumption from onboard systems as well as propulsion is increasingly important. By reducing fuel consumption, operating expenses may be lowered and emissions reduced. Conventional propulsion and energy generation systems on passenger vessels have a high degree of average energy consumption. On diesel-powered passenger vessels that run according to a fixed timetable, fuel consumption typically equates to 35-40% of ship operating expenses (Hoeller, 2014). Achieving significant savings in such energy use therefore is not only environmentally friendly, but also economically worthwhile.

In order to achieve these goals, Shiptec AG has completed several studies and detailed measurements, in order to record the differences between conventional diesel systems and a new (at this time potential) hybrid system that makes use of energy storage (batteries) in order to break peak demand, more evenly distribute energy generation requirements and therefore save fuel. The evaluation compares existing ships fitted with conventional diesel engines to a new vessel equipped with the hybrid system.

1.1 THE NEW VESSEL

The customer's specification called for a vessel that "sets a new standard with regards to comfort, operational flexibility and energy consumption".

The client placed great importance on a vessel that is quiet and has as little vibrations emissions as possible. Generally, these demands are now commonplace for modern passenger ships, as travelers are to enjoy tranquility as much as possible. The intended operational duties of the new ship are regularly scheduled, timetabled lake cruises (75%) as well as specially themed events (15%) and private charters (10%) (Kluser, 2013). In order to meet these varied operational requirements, a new and highly efficient energy and propulsion system was needed. For example fast response time to increased power and propeller speed. High availability and reliability of the propulsion system and the onboard power supply are also essential, as are high efficiency, low TOC and safe crash-stop behaviour (<3.5 x -DWL).

<u>Technical data:</u>
LOA: 63.2 m
DWL: 60.3 m
BOA: 13.2 m
BWL: 10.1 m
Displacement: ca. 400 t
Passenger capacity: 1100
Crew: 4
Shaft power: 2 x 585 kW

2. ANALYSIS OF THE OPERATING PROFILE

Already during the conceptual design stage for the new energy and propulsion system, an emphasis was placed on risk reduction. Benchmarks such as ship type, ship deployment and mode of operation always depend greatly on the individual owner and the crew. Other external factors and the environment of the operating area also have to be factored in. These parameters make it difficult to compare current with previous projects. Often ship owners are not willing to tread new ground and trust emerging technologies, due to a lack of previous experience or uncertainties about future cost or reliability. In some operating environments, the close integration of water with land transport also has significant influence on the design parameters, as do weather conditions in extreme environments. In these cases, "tried and tested" technologies usually win out.

When analysing all of the potential risks, it quickly becomes clear that a complex and fully integrated energy and propulsion system can only be designed according to accurate measurements of similar operating profiles combined with detailed simulation and prognosis about the expected deployment of the ship. Typically, this kind of due diligence is not practiced during the design and production of inland vessels. Instead, new ships are often built around a spec that includes generous power reserves. These reserves, however, come at the cost of operating efficiency and operating expense. Because the greatest savings in energy consumption and maintenance cost are achieved by downsizing, it is worth knowing exactly to what extent one actually needs to downsize engines and gensets. Only by completing this detailed analysis and simulation process, will one obtain the necessary data to specify the degree of downsizing that is necessary to yield financial savings while limiting technical risk.

2.1 OPERATING PROFILE

At the start of the campaign, exact measurement systems and data recording processes first had to be developed, in order to reliably measure actual power requirements for the propulsion as well as the hotel load. The measurement system had to be dynamic and adaptable to the different operating conditions.

An existing ship was chosen as a substitute for the future vessel and as the testbed case, the regular timetable route from Lucerne to Flüelen was selected on Lake Lucerne in Switzerland. This regularly scheduled, timetabled service is about 93 km/51 Sm (Achermann, 2014):

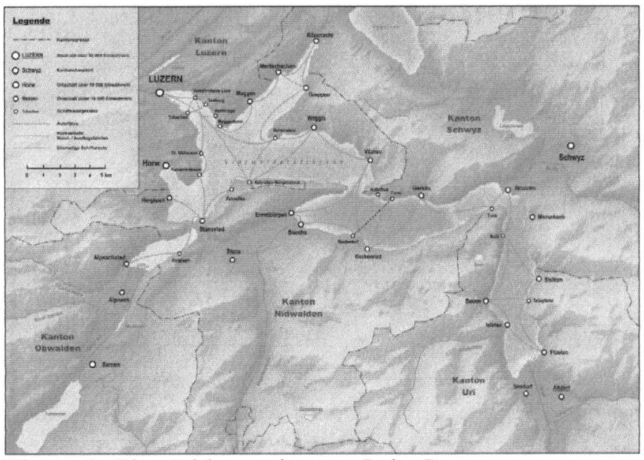

Figure 1: Timetable services on Lake Lucerne

Using this measurement system and corresponding methodology, it is possible to establish an operating cycle that should be similar to that of the future ship and therefore can be calibrated with predictions from CFD calculations and towing tank (with future hull design and weight) (Van Heerd, 2014). Therefore, this overall picture provides a reliable forecast of the dynamic power requirements for the new ship.

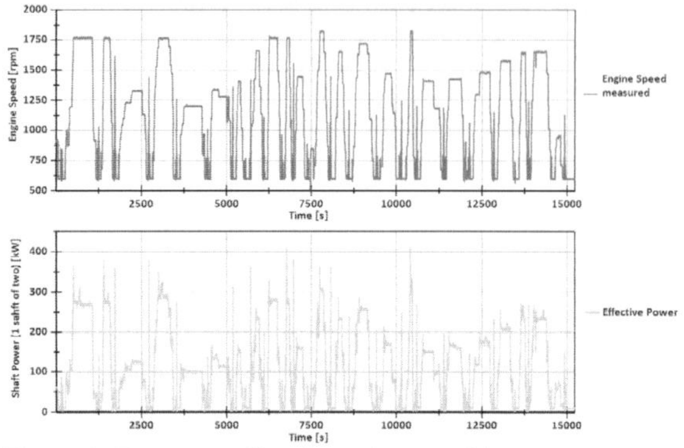

Figure 2: Data recording from the test ship

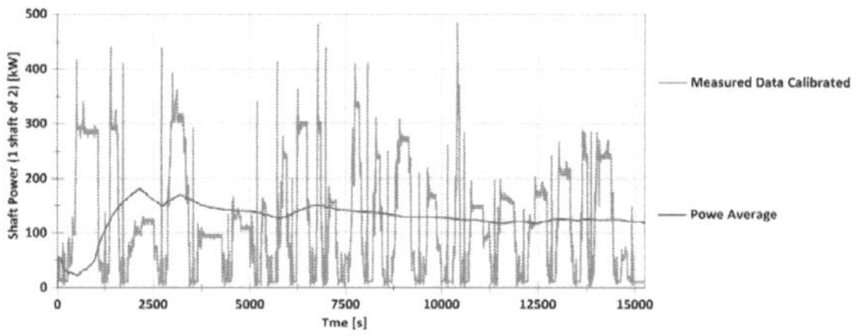

Figure 3: Load case calibrated with towing tank test results

As the engine speed is of lesser importance (assuming that this value is the same for the future ship as the test vessel), only the effective power requirement has to be considered in further steps.

As can be seen in figure 3, the calculated average power requirement (red line) will be quite small in relation to projected peak power. This illustrates the direct potential of a hybrid solution, since conventional propulsion systems must always be designed for maximum peak power during manoeuvers, and the motorization is always oversized when in normal operating mode. Correspondingly, the over dimensioned engines are usually running in an inefficient operation point most of the time. If such over-dimensioned peak power can be broken with the aid of an electric booster, which draws its energy from batteries that are charged by the surplus power of the engine during low (constant) power conditions, effective power compensation and thereby a successful downsizing of the engine can be achieved.

Based on these considerations, the basic concept for the prototype was developed as a parallel hybrid. In this system, a relatively small electric motor is used as a booster to assist the engine during dynamic processes or in continuous operation at constant speed as a shaft generator, while still also generating energy to charge the buffer batteries and the onboard electrical system (see 2.2).

It should be noted that following this hybrid concept the electrical efficiency losses due to the use of inverters occur only on the power generation side during stationary driving (constant ship speed), but not in actual propulsion.

As well at round trip cruises or ferry operations, differences in performance do occur during the journey, but not with so many transient processes during a short period of time, as is the case under timetable services. This means that round-trip cruises or ferry services with precisely defined load conditions are actually well suited for serial hybrids. Thereby, the energy-converting engines (generators) can be designed in such a way that the generator operates at its optimum point at a pre-defined vessel speed, while another generator or the combination of several more generators operate optimally at different vessel speeds.

2.2 INTEGRATION OF ONBOARD ENERGY CONSUMERS

As a next step, the system is supplemented by the integration of predicted, partially dynamic onboard energy consumption.

Observing conventional energy systems, one notices that the gensets often have a very poor efficiency because they must be designed for a theoretical peak load, while the average load is much lower. The advantage of a parallel hybrid system is that surplus power from the above mentioned shaft generators can be used to feed the hotel load, as long as the extra power is not needed to recharge the batteries (considering the power range between the propeller curve and the full-load curve of the engine). As part of this project, it has become apparent that as long as the energy management system and the battery capacity are correctly aligned, any shortfall in the provision of onboard energy can be compensated by the batteries. This also works if the shaft generator can only deliver energy sporadically, which is usually the case when ships running regular timetabled services during docking times at the different piers.

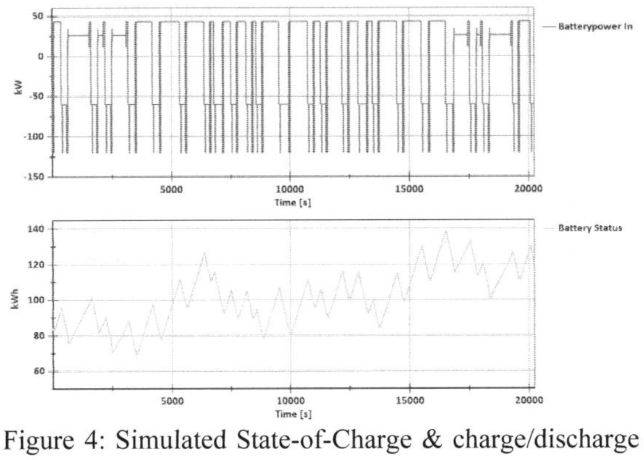

Figure 4: Simulated State-of-Charge & charge/discharge power

In order to achieve proper system design, all of these considerations, including actual measurements as well as future consumption predictions, must be bundled together, so that the theoretical as well as dynamic behaviour of the future batteries can be depicted through a simulation (see figure 4). The simulation results then form the basis for the actual systems design process and for selecting the right components and closed loop processes.

2.3 SIMULATION PROGRAM

At the beginning of the project, the calculations and forecasts mentioned above were based on a discrete calculation method, but eventually it became impossible to process the high data volume with this kind of methodology. Therefore, Shiptec developed a numerical simulation program based on real measured data. It provides simple and rational variable variations for initial conditions as well as operational parameters. This makes it possible to quickly create an initial design structure before completing iterative refinement at a later stage.

When used in combination with measurement systems and typical shipbuilding processes, including relevant calculations and trials, this simulation system will yield reliable predictions with acceptable risks.

3. SYSTEM ARCHITECTURE AND SYSTEM DESIGN

3.1 SYSTEMS ARCHITECTURE FOR THE 2 PROPELLER PROTOTYPE SHIP

The following figure 5 shows the systems architecture of the new ship. In addition to the actual propulsion system with buffer batteries, the vessel is also fitted with a bow thruster.

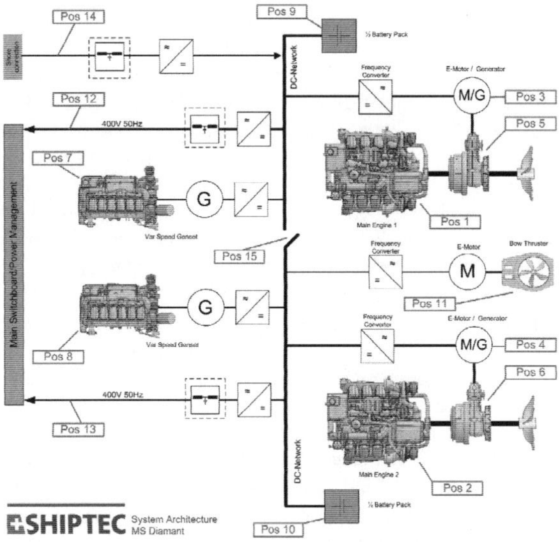

Figure 5: System architecture of the prototype vessel

3.2 COMPONENT SPECIFICATION

- Propeller shaft power 2 x 585 kW (2 x 405 kW engine [Pos 1/2], 2 x 180 kW electric motor/generator parallel [Pos 3/4])
- Boost power for main engine: 2 x 700 Nm from zero to 1200 rpm (engine rpm) through the electric motor [Pos 3/4]
- Power of shaft generator: 2 x 60 kW [Pos 3/4]
- Hybrid Gearbox [Pos 5/6]
- Peak load generators for increased energy consumption (e.g.: advanced catering): 2 x 180 kW electric output [Pos 7/8]
- Buffer battery capacity [Pos 9/10]: 2 x 42.2 kW/h: designed in such a way that the peak load generators do not have to be switched on during standstill periods when the ship calls at piers during timetable operation. There is sufficient energy to provide boost power during engine acceleration
- Bow thruster: 1 x 125 kW [Pos 11]
- Max. propeller shaft power in electric mode: 2 x 150 kW [Pos 3/4]
- Power network supply: 2 x 100 kW (partially redundant) [Pos 12/13]
- Plugin for onshore power supply (extra power while the ship is being prepared or for over-night charging of the batteries) 1 x 100 kW [Pos 14]
- System separator, so that in case of an emergency the entire system may be operated as two separate systems (starboard and portside) [Pos 15]

3.3 SYSTEM FUNCTIONALITY

Based on the power requirements of the propulsion, the hotel load and the batteries, the energy system is able to recognize the current load of the different engines and can alternately load, engage or disengage them. Based on the dynamic power requirements, the energy system also recognizes whether the electric motor/generator on the propeller shaft should automatically be switched into boost mode or vice versa. Additionally, the fully integrated power management system also regulates energy supply between batteries, shaft generator and in the worst case the peak load generator, so as to achieve optimal and reliable power supply for all energy consumers. In case of significant power demand, the system is supported by automatic or manual heavy consumer requests. This means that power requests from kitchen equipment are automatically recognized, whereas the bow thruster can be manually allocated by the skipper. For critical nautical manoeuvers, such as difficult traffic situations or ship behaviour in a storm, the skipper can switch on an additional power function, which blocks the shaft generator function and instead reassigns the electric motor to work only as a booster for the power transmission. In such an instance, the two peak load generators are switched on and are used to supply the entire hotel grid, including the bow thruster and the booster. In case of a critical situation, the skipper therefore has the combined power of the entire system at his disposal.

4. PRACTICAL FUNCTIONALITY

4.1 RESULTS FROM SEA TRIALS

4.1 (a) Constant speed trials

These trials are intended to verify the power consumption of the propellers in function of the propeller shaft speed. As such, these recordings provide essential base data from which other parameters are then derived. With a parallel hybrid system, this knowledge is all the more important, as the propeller shaft values will have effects on requirements and performance of the other equipment, including engine, shaft generator and booster.

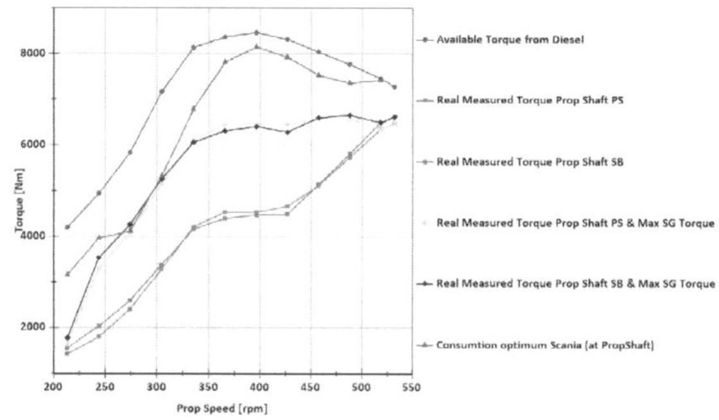

Figure 6: Results constant speed trials

Figure 6 shows the effective propeller curve. Additionally, the theoretical optimum consumption value of the main engine is shown in dark green (Scania AB, 2014). From this chart, it is apparent that especially in the higher speed range less specific fuel consumption may be achieved, than was originally anticipated during the very conservative parameterization process. The initial draft parameterization primarily sought to achieve enough of a safety margin between the standard and full load curve. This gap establishes the dynamic potential for rapid speed changes during relevant operational circumstances.

Since the completion of comprehensive commissioning trials and based on initial operating experience from actual revenue service, it has been possible to adjust characteristic curves. This limits the maximum power deployment of the shaft generator and therefore allows the system to be further optimized, achieving even more efficient values between 350 and 520 rpm in relevant areas.

4.1 (b) Pier manoeuvers

Pier manoeuvers are the departure and arrival of the ship from different piers. In regular timetabled operations, this process usually repeats itself every 5 to 10 mins (Achermann, 2014).

In addition to constant speed trials, which were designed to test the boundaries of the system, much emphasis was placed on the dynamic evaluation of typical operational situations. This was intended to parameterize the basic design concept as well as verifying the dynamic and potentials of the propulsion system, while verifying and refining energy distribution to the hotel grid.

Figure 7 below depicts the typical sequence of a pier manoeuver (sampling rate 0.02s):

- a Cruising speed
- b Speed reduction
- c Propulsion disengagement, speed release
- d Astern propulsion burst
- e Stop (time required to reach standstill)
- f Gradual acceleration
- g Resumption of cruising speed

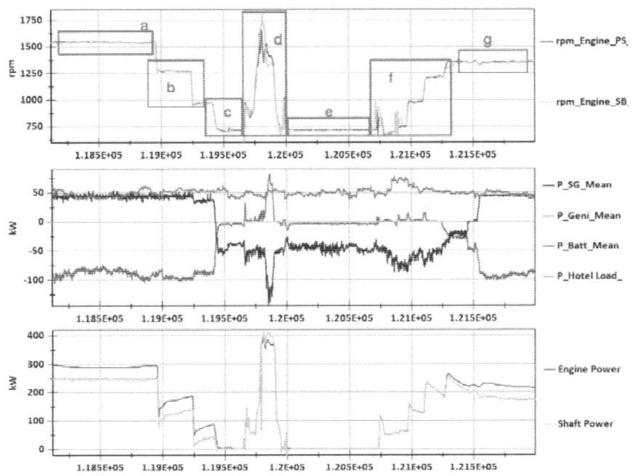

Figure 7: Pier manoeuver

The upper part of the figure shows the effective engine speed during the manoeuvre. Since both engines can be controlled independently, differences between the portside and starboard system are noticeable here. They are caused by the prevailing environmental conditions at the pier (such as wind, water flow and physical space).

The third part of the graphic illustrates the situation on the propeller shaft as well as on the main engine. As such, the green curve represents the effective power on the propeller shaft and the red curve displays the power of the engine. Two effects become apparent:

In *a*, *b* and *g* it can be seen how the shaft generator is powered and supplies the vessel with electric current plus battery charging. The graph in the middle highlights this with the red line recording shaft generator power and the blue line indicating battery charging power. In areas c, d, e and f, the shaft generator is not active, instead the board network (as shown in the magenta curve) is being supplied by the batteries. The corresponding negative battery charge is shown by the blue curve. In area d and somewhat also f, the booster comes into play and has been supplied by the batteries, as seen with the red lines marking peak periods. Once this value becomes positive, actual mechanical power is released. The energy for the booster is supplied by the batteries (middle chart showing blue power peaks towards the bottom). As the chart further illustrates, genset power remained "0" during the entire process, without any need for the genset to kick in during the pier manoeuver and the batteries taking care of all energy supply to the hotel load instead.

4.1 (c) Crash stop manoeuver

As already mentioned, every vessel in Switzerland must satisfy specific crash stop benchmarks in order to receive certification (Schweizerische Eidgenossenschaft, 1994). The necessary test is rather easy to execute and reproduce, but it does place big demands on the system in terms of fast and precise reaction time and requires very high torque on the propeller shaft at very low rpms.

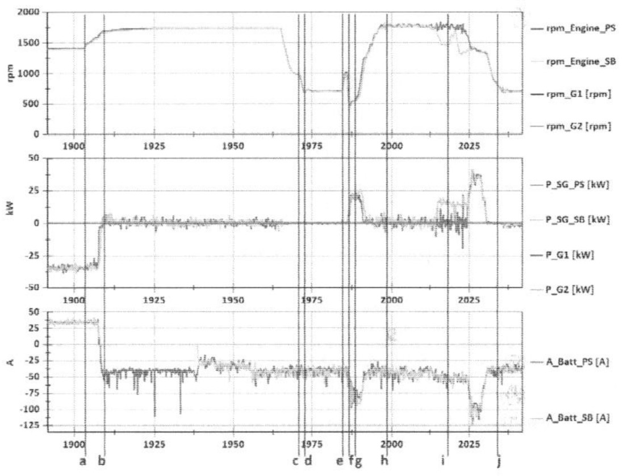

Figure 8: Crash stop manoeuver

From "full ahead", the control lever is pulled back to "full astern". In these instances, the system implements a pre-defined process, which is characterized by a maximum permissible time and distance until the vessel has reached a complete stop. Figure 8 above shows this process, with a sampling rate of 0.02 second:

- a Acceleration to maximum speed
- b Achievement of max. speed
- c Reduction of propeller speed after the retraction of the control lever (equivalent to idle speed of the engine)
- d Disengaging the propeller shaft and braking the shaft with the shaft brake
- e Acceleration of the engine prior to engagement of the propulsion
- f Engage propulsion
- g Acceleration of the engine with the propeller rotating backwards under full load

h Propeller slippage (air)
i Propeller grip with full torque
j Ship has come to a complete stop

The upper part of the diagram shows the speed of the main engines as well as at the gensets. It is noticeable that the gensets do not start during the entire process. The energy for the hotel load as well as the booster is only drawn from the batteries.

Shown also at the lowest chart from figure 8. The negative battery charge current, which is displayed in red and green, represents discharge. The chart in the middle illustrates the power of the shaft generators. It shows *a-b* the acceleration to maximum cruising speed, as well as shaft generator power goes to 0 kW, in order to provide the propeller with the entire power output from the main engines. Position *f* shows how the shaft generator becomes a booster during this process, thereby supporting the main engine with maximum boost torque. Position *i* records propeller grip at full torque.

If the downsized prime mover (engine) would have to make due without a booster, it would not be able to produce peak output for either rapid acceleration or adequate propeller grip at full torque. The results described above demonstrate that the main requirement for fast torque delivery to the propulsion system has been fulfilled, even though the main engine is purposefully small, for the sake of efficiency and economy.

Just to compare: In order to achieve such performance with a conventional engine drive, an engine with about 50-100% more stroke volume would have to be selected. This, however, would lead to a massive underloading of the engine and contribute to poor efficiency. Moreover, the maintenance costs would be significantly higher over the entire life cycle of this larger and therefore more expensive engine (Durrer, 2014).

4.1 (d) Fast stopping manoeuver at high hotel loads

Generally, this manoeuver does not differ from the above-mentioned crash stop procedure; merely the initial speed is lower. This means that the manoeuver does not take place through previously reduced shaft generator power, but occurs directly from the shaft generator function. This feature also makes the system more agile. To create an extra challenge for these trials, the load requirements of the onboard network (catering services working at peak) were purposefully increased, so that the maximum theoretical power of the batteries would no longer be sufficient to support the entire electrical system (fourth curve in figure 9). In this extreme situation, the gensets therefore has to automatically start and deliver extra power.

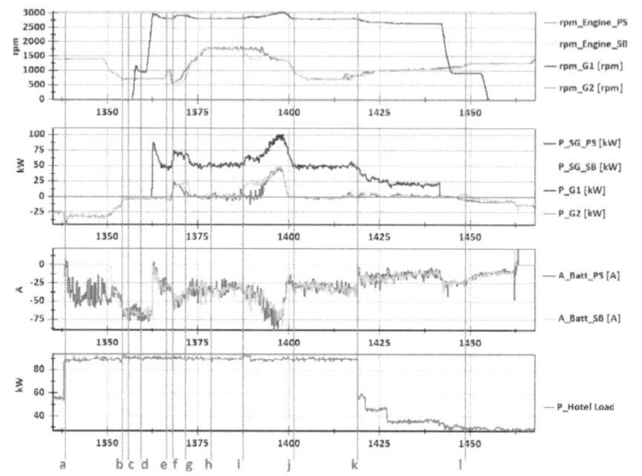

Figure 9: Fast stopping manoeuver at high hotel loads

Procedure (sampling rate 0.02s):

a Cruising speed, load entrance from the hotel load
b Reduction of propeller speed after the retraction of the control lever (equivalent to the idle speed of the engine)
c Disengaging the propeller shaft and braking the shaft with the shaft brake
d Start genset
e Engine acceleration prior to engagement of the propulsion
f Engage propulsion
g Acceleration of the engine with the propeller rotating backwards under full load
h Propeller slippage (air)
i Propeller grip with full torque
j Ship has come to a complete stop
k Reduction of hotel load power
l Stop genset

After switching on the increased network load, it can be noticed how the batteries take over the load together with the shaft generator *a*. In case propeller speed is now reduced, the potential power output of the batteries is no longer sufficient to supply the entire electrical system, due to the reduction in power output from the shaft generator. In order to still have adequate reserves for a potential peak in hotel load requirements, a genset is activated. This helps regulate the required energy supply for the hotel load and provide boost power, if required *g* and *i*. Once peak load has been reduced after the manoeuver (test condition), the genset automatically switched off. These behaviours ensure that the entire energy and propulsion system can function as a whole (as it is required in operation). This kind of full systems integration allows the genset to start for short boost periods during peak demand, before switching off, once the intensive energy requirements have subsided.

4.2 INITIAL RESULTS FROM DEPLOYMENT ON REVENUE SERVICES

In addition to the sea trials, initial results from regular operations in actual service have become available for review and comparison with the originally projected energy savings.

Timetable operation 11-20 (Luzern – Flüelen – Luzern) (Achermann, 2014)

The diagrams below show the operating conditions during a timetabled cruise from Lucerne to Flüelen and back. During the measurement process, a sample rate of 0.02 s was used. This data can be compared to the measurements and simulation records that were originally collected at the beginning of the project. Figure 10 below corresponds to an entire timetabled cruise with no less than 24 calling points. All of these pier manoeuvers correspond to the processes that have already been detailed in section 5.1.2. For the sake of this general overview, the individual manoeuvers are less important, but rather the entire power distribution cycle and energy flow should be analyzed as a whole. During such timetable cruises, energy demand includes peak periods such as hotel loads, propulsion loads and the bow thruster come into simultaneous use. Figures 10 and 11 both document the exact same cruise. In figure 10, the focus is on speeds of the propulsion system, the comparison between power flows as well at the power on the engine and the propeller. Figure 11, represents the condition at the buffer batteries and the corresponding power flows.

The second curve of figure 10 shows that the system has the required variability to switch between different power sources to cover varying degrees of energy requirements. This so called "load sharing" is easy to follow visually. While the entire power on the hotel load is illustrated in magenta colour and remains quite consistent for the full duration of the timetabled cruise, the energy sources respond differently, mainly depending on the rpms of the main engine. At one point, the batteries are being recharged and the hotel load completely supplied by the shaft generators (typically during cruise speed between different stations), while at a different point e.g. during pier manoeuvers the electric motors kick in as boosters (red line in the positive area). Later, the hotel load draws its energy from the batteries (blue line in the negative area). The light blue lines represents shore connection energy supplied to the ship. Occasionally, peak power consumption and a momentary boost from a genset is seen in green. The gensets starts, when the bow thruster is required, on top of high demands placed on the hotel load and the catering equipment. The third curve of figure 10 shows the effective power at the main engine (red) as well as the propeller shaft (green). The noticeable difference between these two curves is due to the shaft generator power (red higher than green) or the corresponding boost effect (green higher than red, though hardly apparent in the chart).

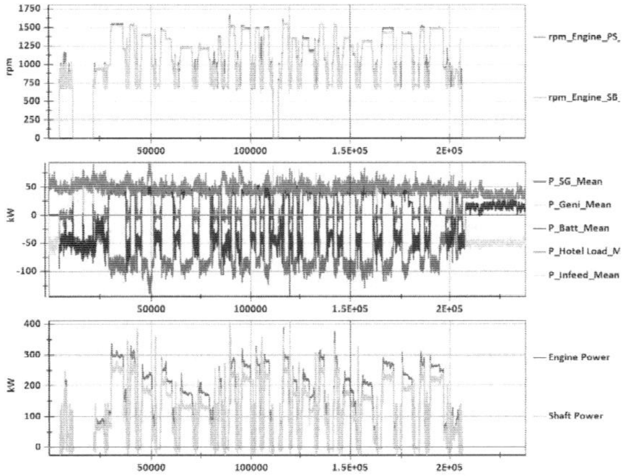

Figure 10: Operation 11-20 Lucerne-Flüelen-Lucerne: Engine speed and power

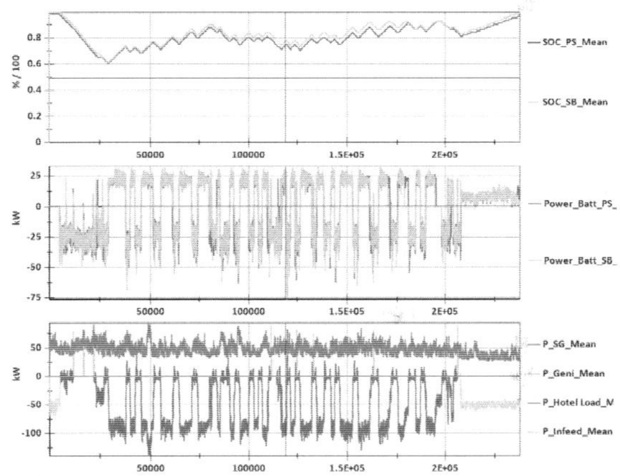

Figure 11: Operation 11-20 Lucerne-Flüelen-Lucerne: Battery SOC and power flow

Figure 11 records mainly battery performance. The first curve displays the state of charge (SOC) during the entire timetable operation. Here it is noticeable, that the specific use of battery capacity and function has been well planned and optimally implemented. During the whole cruise, SOC is more or less constant, owing to evenly distributed energy requirements and supply balance.

The second curve illustrates the re-charging and discharging of both battery packs. Especially during the discharging process, there are no significant, negative peaks. Even though the technical specification for the battery discharging capacity is roughly 70 kW per pack, the effective performance should not be that high, so as to allow for enough energy reserves in case of a sudden, worst-case scenario with significant power demands. If that kind of reserve energy would not be available, the gensets would not be capable to react quickly enough with additional power supply, which in turn would raise the risk of a blackout aboard the ship. Because the recharging capability of the batteries is limited electronically, no positive excess can be expected. The third curve range in figure 11 is analog to the second curve series in figure 10, with the exception of battery flow, which has been extracted in the second series.

Finally, the third figure to document ship performance during regular, timetabled operation (figure 12) records the propeller shaft energy demand and the energy output from the main engine. It is therefore possible to notice the difference between the effectively utilized energy on the propeller shaft (green), which has been supplied by the main diesel motor (red).

These recordings show that the load of the main engine is about 20% better than aboard a comparable vessel with a conventional diesel drive. This higher load is mainly possible due to the power taken by the shaft generator. Thanks to this effect, the main engine gains a better degree of efficiency, while the hotel load does not require other purely dedicated generators that run at even less efficiency.

If one compares the 20% higher load of the main engines with the originally predicted values (see section 4.1 (a)), one notices that even greater potential still exists for future optimization. Using constant monitoring and collecting of operational data, such improvements can be implemented in future. Indeed, more optimization work on the new vessel is already planned for winter 2017-18.

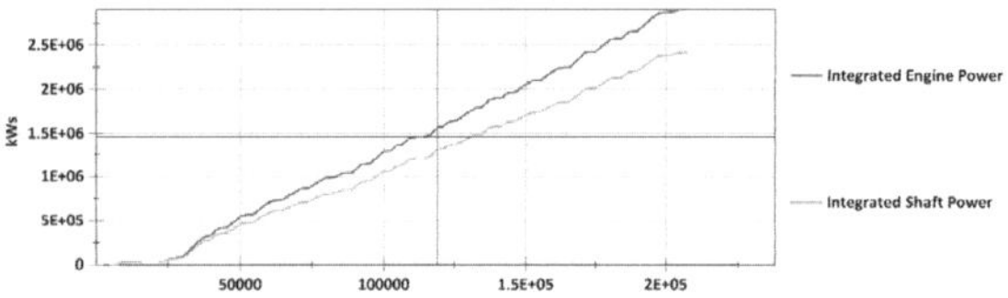

Figure 12: Operation 11-20 Lucerne-Flüelen-Lucerne: Energy

5. CONCLUSION

5.1 FUEL CONSUMPTION ANALYSIS

Because it is impossible to compare the operational parameters of different event cruises with each other, the measured fuel consumption is limited to data from timetabled cruises. These account for roughly 75% of the newt ship's completed mileage (Kluser, 2013).

In order to properly evaluate fuel consumption, the integrated consumption data from the main engines was used, by cross referencing it with the fuel tank gauge before and after every timetable cruise.

To establish a calculation basis for the fuel consumption levels based on the energy needs of the hotel load, the electrical power requirements for a full-length timetable cruise was integrated and multiplied by an average fuel consumption value of 193 g/kWh. This value is basically lower than that of a genset running in his optimum. Therefore, optimal consumption values should be classified based on electric output, while energy usage for the propulsion system should be seen as worst case.

In order to evaluate the overall success of the system, the differentiation between energy consumption from the hotel load on one side and propulsion energy on the other, is indispensable because comparable data is only available for these two categories. At the same time, a certain degree of fault tolerance must be given to this evaluation method.

In addition to this differentiation between hotel load and propulsion, one needs to factor in different ship sizes when evaluating energy consumption of propulsion. In that regard, ship weight is a helpful benchmark. Based on past experience, when ship type, build size, speed capability and driving style are all rather similar, fuel consumption will mainly depend linear on vessel weight. The following consumption values are therefore divided by 100 t of ship weight, so as to have comparable benchmarks per propelled ton.

Table 1: Consumption values

	CONSUMPTION TOTAL	CONSUMTION HOTEL LOAD	CONSUMTION PROPULSION	WEIGHT HALF-LOADED	CONSUMPTION PER 100 T OF VESSEL (HALF-LOADED)
NEW VESSEL OPERATION 11-20 10.05.17	4.08 L/km	0.84 L/km	3.24 L/km	393 t	0.82 L/km
NEW VESSEL OPERATION 11-20 04.06.17	4.12 L/km	0.82 L/km	3.3 L/km	393 t	0.84 L/km
MS EUROPA OPERATION AVERAGE 2016 [8]	**3.9 L/km**	**0.6 L/km**	**3.3 L/km**	**275 t**	**1.2 L/km**
MS WALDSTÄTTER OPERATION AVERAGE 2016 [8]	3.7 L/km	0.4 L/km	3.3 L/km	300 t	1.1 L/km

By glancing at the different fuel consumption values listed in table 1, the following conclusion becomes relevant:

- The new vessel has higher energy consumption for its hotel load. This circumstance is due to the electrically operated kitchen and a powerful air-conditioning system, as well as generally more "energy hungry" devices installed on the ship. If one considers these circumstances from a differentiated point of view, one notices that the extra fuel consumption is actually not that much higher, given all of the extra comfort and amenities that the vessel offers. For instance, on the other Lake Lucerne ships the average fuel consumption statistic for electrical energy generation is 0.7 L/km without air-conditioning and with ship kitchens that mainly operate on butane gas (Strohammer, 2017). Two circumstances have therefore contributed to a recent reduction in energy consumption:
  - Specific measures that were first identified through a scientific research project are starting

to yield results, as there has been a successful reduction in energy demand from secondary consumers such as heating and ventilation as well as lighting.

- The benchmark of 193 g/kWh average diesel fuel consumption for electricity generation is rather low and a further adjustment of this value would also result in less fuel consumption for propulsion purposes.

With a high degree of certainty, both of these factors have contributed to this estimated low fuel consumption.

- The absolute consumption per kilometre is actually in the same range as for comparable vessels, even though the new ship has 100 t more weight and capacity for 400 extra passengers.
- If fuel consumption per kilometre is divided through 100 t weight of ships, including 50% passenger capacity, the new ship suddenly has average fuel consumption of 25-30% less than the ships with conventional diesel propulsion. Compared to the originally calculated fuel savings of 14-17% (as verified by DNV-GL Maritime Advisory as well as the Swiss Federal Institute of Technology - ETH Zürich [Pape, 2014; Ritter & Onder, 2014]), it is possible to conclude that the original project expectations have been exceeded and other statistical methods (such as fuel consumption per passenger capacity) would reconfirm this.

More precise statements and technical data will become available after joint research project with the Lucerne University of Applied Sciences has concluded in late 2019. Based on the data recorded in that campaign, it will be possible to verify the assumptions as well as place them into a broader context.

5.2 MAINTENANCE COSTS

Another positive effect of the parallel hybrid energy and propulsion system is the significant cost reductions that it allows in terms of maintenance, because of its smaller, downsized components. It is still difficult to exactly quantify this aspect, as until now no significant or unexpected maintenance work has had to be completed on the new ship, since its official launch in May 2017.

But initial findings have shown, that at the current pace of operations (2500 service hours per year), regularly scheduled maintenance work for the generators will only be required every two years instead of twice a year. Additionally, the smaller main engines only require about 65% of the maintenance costs of larger, more powerful engines, as they are fitted on conventional motor vessels.

5.3 INVESTMENT COSTS (BASIS FOLLOW UP ACCOUNTING)

A parallel hybrid system, as it is described it in this paper, does have higher up-front investment costs than a conventional energy and propulsion system. This is mainly due to the components, such hybrid gearbox, battery packs, electrical motor and closed loop regulation. All of these "off the shelf" components are required to complete the system and make it work. Follow-up accounting has shown that including the somewhat more demanding installation process, the expected in-vestment costs for a parallel hybrid system are about 27% higher than for a conventional system.

Taking into consideration the lower operating costs and reduced maintenance expenses of the used parallel hybrid system, the 27% higher up-front investment will have paid for itself after only 3 ½ years at 2500 operational hours per year. This is calculated without interest rates, but takes total cost of ownership (TCO) into account.

6. REFERENCES

1. HOELLER, W., '*Kalkulation Betriebskosten MS Bürgenstock*', SGV AG, 2014.
2. KLUSER, T., '*MS 2017 Anforderungsprofil'*, SGV AG, 2013.
3. ACHERMANN, M., '*SGV AG Fahrplan Sommer 2014'*, www.lakelucerne.ch, 2014.
4. VAN HEERD, J., '*Passenger vessel for Lake Lucerne (60m); calm water model tests*', Marin Report No. 27202-2-DT, 2014.
5. SCANIA AB, '*BSFC Map DI13 070M 405kW*', Tech. Documentation Scania AB, 2014.
6. SCHWEIZERISCHE EIDGENOSSENSCHAFT, '*Verordnung über den Bau und Betrieb von Schiffen und Anlagen öffentlicher Schifffahrtsunternehmen*', www.admin.ch, 1994.
7. DURRER, L., '*Schlussbericht MS Schwyz Sanierung 2012/2013*', Shiptec AG, 2014.
8. STROHHAMMER, G., '*Treibstoffverbrauch Flotte SGV ab 2013*', SGV AG, 2017.
9. PAPE, B., 'MS 2017 – *Konzeptvergleich zweier Antriebssysteme*', DNV-GL Report, 2014.
10. RITTER, A. & ONDER, C. DR., '*Energetische Betrachtung des hybriden Fahrgastschiffes MS 2017*', Swiss Federal Institute of Technology Zurich, 2014.

7. AUTHORS BIOGRAPHY

Martin Einsiedler holds the current position of Head of Naval architecture and Engineering at Shiptec AG. He is with his entire team responsible for engineering, researches and development as well business development in different markets for marine engineering and ship design. His previous experience includes amongst others energy system development for smart grid applications (standalone application for small power plants) as well the corresponding combustion development for diesel engines in combination with the production of electrical power and propulsion in vessel and vehicle applications.

DESIGN-SIDE INNOVATION TO MINIMIZE THE ENVIRONMENTAL FOOTPRINT OF A RO/PAX FERRY

D M MacPherson, HydroComp, Inc., USA
I Bakas, NAP Engineering P.A., Greece

SUMMARY

The EU co-funded POSEIDON MED II Deliverable 14.4, which initiated this work, calls for an LNG-fueled Ro/Pax ferry design, which included a mandate from the client ("Blue Star Ferries", also a program partner) to investigate innovations focusing on the "footprint" of the vessel with regard to the ocean environment and fuel efficiency. The mandate was for innovations in both operation and design.

This presentation will discuss the design-side investigations and their outcomes. New calculations proved vital to the success of the project, including use of a new computationally-efficient resistance prediction method that assists in the search for drag reduction from hull form changes, multi-mode duty profile optimization, and propulsion analysis prediction LNG and dual-fuel consumption indices. The presentation will also address new environment-focused analyses that were developed especially for this project, including reduction in the ship's carbon footprint with full operational profile assessment of CO2 emissions.

1. INTRODUCTION

There is little need to reiterate the significance that greenhouse gases (GHG), such as carbon dioxide (CO2), has become as a naval architectural design driver. Not only are regulatory efforts being enforced for limits on GHG production, owners and other stakeholders are voluntarily instructing designers to use existing and innovative efforts to reduce their "carbon footprint". Such as mandate was the genesis for the Ro/Pax project described herein.

This project called for an assessment of total-voyage GHG production and fuel consumption, with the design objective to be minimization of both fuel and CO2. To allow for minimization of CO2 production, Blue Star Ferries "Statement of Requirements" (SOR) requested consideration of alternative lower-carbon fuels, specifically LNG. Further, the authors delivered additional reduction of fuel consumption and CO2 production via improvements in the ship sectional area curve that were exposed using a novel and cost-effective wave-making resistance prediction code.

Two designs are compared in this paper: a) the original basis hull geometry with a conventional marine diesel engine; and b) the original hull with proposed modified sectional area curve at the stern with a dual-fuel (LNG/MDO) main engine. The modified Ro/Pax hull with dual-fuel engine offers significant savings in both mass of fuel consumed (-23%) and CO2 production (-34%), leading to annual savings of 4900 t fuel and reduction of 3100 t CO2 production.

2. PRINCIPAL CHARACTERISTICS AND VOYAGE PROFILE

The Ro/Pax design is a compact twin-screw vessel constrained to be nominally 140 m length on WL, 24.5 m beam on WL, and 6.5 m operational draft. Propeller diameter was restricted to 4600 mm. Initial hull geometry was provided for the study by Blue Star Ferries following development by principal designer "NAP Engineering", another program partner.

The client specified a representative voyage profile for three sets of departure/return trips per week with annual duty of 48 weeks. The total distance spans 354.6 nautical miles. A collection of individual transits and movements were collected and summarized into three representative "modes":

Slow	8 kts	7.1 nm	0.9 hrs	MDO
Moderate	18 kts	6.7 nm	0.4 hrs	MDO
Cruise	22 kts	340.8 nm	15.5 hrs	MDO & LNG

The condition of normal ship operation for the propulsion analysis is:

- Ship with appendages in a clean condition. (No added roughness.)
- Head wind and seas corresponding to Sea State 3. (No shallow water correction.)
- Bare-hull resistance design margin of 8%.
- Design speed of 22 knots.

3. COMPUTATIONAL ANALYSIS METHODS

All calculations for the prediction of resistance, propulsion, voyage modes and summaries, fuel consumption, and CO2 production were conducted using the HydroComp NavCad® software (a resistance and propulsion system analysis tool with 600+ commercial users in more than 40 countries). New modules were developed by HydroComp for NavCad to support the application of dual-fuel operation, including user specification of fuel properties, prediction of concurrent multiple fuel rates, and for CO2 production. Much of the new development work was sponsored by NAP Engineering.

3.1 RESISTANCE PREDICTION

The NavCad software has some 40 parametric methods available for prediction of bare-hull resistance. The "Method Export" ranking system provides design guidance for the proper selection of a method for the given speed range, hull parameters, and local hull form details (such as support for a bulbous bow or immersed transom). Parametric prediction methods are often called [1D] methods, as they use single-dimensional "descriptors" of a hydrodynamically-relevant hull form characteristic. The collection of these well-known parameters (length, beam, displacement volume, etc.) and their non-dimensional variants (L/B, CP, etc.) describe the "bounding" characteristics of the immersed volume. The parameters used for the prismatic coefficient (CP) are shown in Figure 1 below.

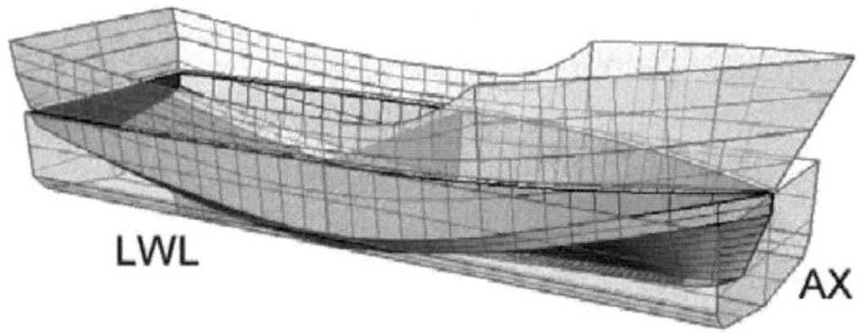

Figure 1: Example [1D] hull form parameters

The strength of parametric methods are many, including an anchor in empirical model tests and the inherent smoothing that comes with statistical regression of many data points within the prediction model. A weakness of parametric methods comes from the necessary simplification of the shape of the immersed volume into individual parameters. Some methods are substantially simplified to a handful of parameters; others utilize an extensive set of shape descriptors.

It is fair to say that all hydrodynamic predictive models simplify the physics. Model tests are still considered the "gold standard" of hydrodynamic prediction, yet they conduct an empirical test on a model of fractional scale in conditions that are not the same as the full-scale ship will encounter. Higher-order computations codes, such as [3D] CFD, create a much more detailed shape of the envelope of the immersed volume, but these calculations are also simplified by grid density and the minute predictors of boundary-layer properties necessary for the prediction.

The authors employed an intermediate analytical method that gives greater detail to the hull form shape descriptors without the financial, time, and resource expense of model testing or higher-order [3D] codes.

Often referred to as a [2D] method, NavCad's "Analytical Distributed Volume Method" (ADVM) is a novel bare-hull resistance prediction method that can be used in place of a parametric method. Inspired by thin-ship wave-making theory, it is a prediction of wave-making drag where the hull form is described by the longitudinal distribution of the immersed volume as described by longitudinal distributions of beam at waterline, sectional area, and the centre of immersed area. (The distributions for the original hull design are shown in Figure 2 below.)

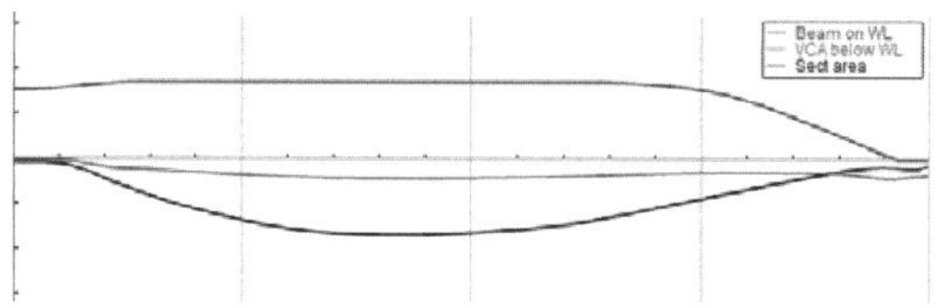

Figure 2: Longitudinal distribution of volume by beam, sectional area, and centre of area immersion

The wave-making resistance prediction component of the ADVM method is based on calculation of the Kelvin wave system of diverging and transverse waves, using a nested integration of the longitudinal distribution of the sectional area characteristics within a wave propagation direction. The basic theory employed in the ADVM is inviscid, so the influence of viscosity on the wave-making components has been considered using a variety of corrections. These corrections are generally to reduce the amplitude of the wave hump-hollow oscillations and to correct the immersed length domain of the Froude number so that the speed positions for the oscillations are correctly predicted. To achieve this end, corrections have been developed and applied for boundary layer displacement thickness addition, transom stern "hollow" extension, flow convergence, dynamic trim, and low speed transom pressure, to name only the most significant. A supplemental outcome of the method is prediction of the far-field wave-patter elevations (often used as a way to communicate energy savings during design optimization). A sample is shown below.

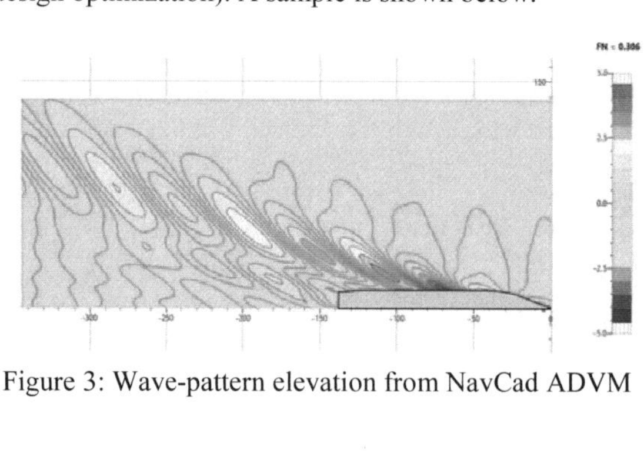

Figure 3: Wave-pattern elevation from NavCad ADVM

Of course, reliable whole-ship prediction of resistance is more than just wave-making drag, so the method also considers scalable viscous properties through a contemporary assessment of form factor and frictional drag. Still, even with the described research and development efforts to properly model and predict viscous effects on the wave-making component, the ADVM (or any other similar [2D] method, for that matter) is not appropriate for all vessels. Therefore, a validation step was conducted before applying the ADVM method to the project for local shape optimization in the hunt for hull form drag reduction.

3.2 ADVM VALIDATION

Validations of resistance prediction methods can be conducting using a variety of basis criteria, including comparisons to model tests or other calculation methods. In both cases, the basis must be public and widely considered as a suitable benchmark for validation of analytical methods. It must also share similarity in the significant hydrodynamic parameters, such as design Froude number (FN), fineness coefficient (CVOL), prismatic coefficient (CP), bulb area ratio (AB/AX), and shape of the sectional area curve. An extensive literature search identified only one benchmark model that fit these objectives: the KRISO ContainerShip (KCS). Its principal dimensional parameters are compared to the Ro/Pax vessel below:

	Design FN	CVOL	CP	AB/AX
Ro/Pax	0.31	6.0	0.62	0.11
KCS	0.28	6.2	0.66	0.09

Comparison of the longitudinal distribution of sectional area, beam, and section VCA are shown below for the the KCS validation model. (Compare to the Ro/Pax distribution from the figure above.) Data for the KCS model hull geometry and test results can be found at: http://www.t2015.nmri.go.jp/kcs.html.

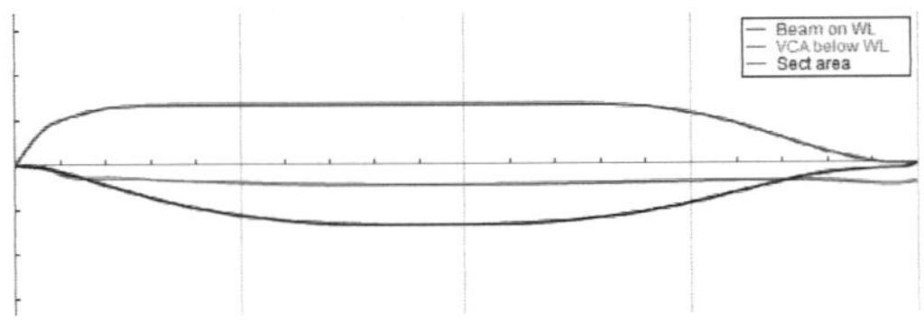

Figure 4: KCS validation model distributions

The KCS model was run in NavCad using the ADVM method, and the results compared to the original model test. To assess curves shapes and magnitudes in the FN range of the Ro/Pax (0.31 FN), the speed range for the ADVM prediction was extended beyond the KCS top speed (0.28 FN). You can see from Figure 5 below that the ADVM captures the drag curve shape of the model test very well, and somewhat overpredicts the resistance. As the over-prediction is a conservative outcome, the ADVM method was deemed an acceptable prediction method for the Ro/Pax hull form and speed range.

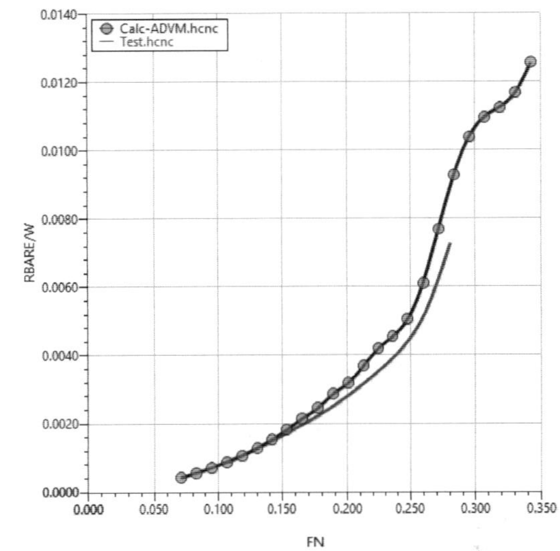

Figure 5: NavCad ADVM validation results

3.3 DEFINITION OF ENGINE FUEL CONSUMPTION

The prediction of fuel rate in NavCad had previously only utilized a *Volumetric fuel rate* (e.g., L/h) for the defined fuel that was to be used in the propulsion analyses. This was sufficient for "always-liquid fuels", but was not sufficient to properly model potentially "non-liquid" fuel types (such as LNG), nor to compare fuels in a dual-fuel setting. To account for this – and for how dual-fuel engine manufacturers typically publish consumption information for their engine models – a number of new data items were added to the NavCad engine model definition for fuel consumption. These include Density, Heating value, Brake Specific Fuel Consumption (BSFC; e.g., g/kW-h), and Brake Specific Energy Consumption (BSEC; e.g., kJ/kW-h). Given the significant differences in fuel density and heating (calorific) value between liquid marine fuels (such as MDO), it is imperative to compare the functional "chemical-energy-to-power" conversion through BSEC.

As one of the important objectives of the owner's SOR was the evaluation of multiple fuel types and dual-fuel operation, new features were developed for NavCad to include:

- Calculation and reporting of the fuel consumption of two user-selected fuels during propulsion analyses. These are indicated as Primary and Secondary fuels.
- Selection from four standard fuel options (plus a defined fuel): Marine Gas Oil (MGO), Marine Diesel Oil (MDO), Heavy Fuel Oil (HFO), and Liquid Natural Gas (LNG). Values for density

and heating value (i.e., lower calorific value) of these four standard fuels are taken from the IMO MEPC 281 (70) reference [1].

- In addition to the calculation of fuel consumption in the steady-state Propulsion analysis, NavCad's Operating modes analysis was updated to allow fuel selection (Primary or Secondary) for individual modes. Plots of fuel use are available within this module for both total volumetric and mass fuel consumption for the voyage, with colour coding for fuel type.

3.4 PREDICTION OF CO2 GREENHOUSE GAS PRODUCTION

One of the important project objectives was to reduce greenhouse gas production, thereby requiring the ability to predict the generation of CO2. This prediction was added to the steady-state Propulsion system analysis (per engine), and also to the Operating modes analysis tool (as the total per mode).

The methodology for prediction of CO2 is quite simple, as it is based on the predicate that CO2 production is directly calculated as a "carbon content" ratio of the mass fuel consumption. In other words, a prescribed portion of a fuel's mass is defined as its unburned "carbon content".

For use with the *Operating modes analysis*, the total CO2 production is calculated as the total for each mode. This allows for unique mode operation to determine if the engine is engaged or not (needed for Idle operation or use of a Compound gearbox), the fuel type specified for the mode, and the mode's total fuel demand. A plot of CO2 production can be shown, whereby the size of the mode's plot marker reflects the relative production. The position of the marker (by power and RPM) indicates the relationship of CO2 production with the engine loading. This plot is a valuable means to communicate the impact of ship and propulsion design decisions to stakeholders..

4. RO/PAX RESISTANCE PREDICTION

A resistance prediction for the Ro/Pax hull was thereby conducted using the ADVM method. The prediction was compared to NavCad's "Confidence plot" (an "optimistic lower-limit" design guide) as shown in Figure 6.

While it is expected that the curve shapes should reasonably match, the distance above the "confidence" lower-limit line would suggested to the authors that the hull is not optimal. A critical review of the hull form was conducted by the authors to identify obvious shape characteristics that might be improved. The stern shoulder (of the sectional area curve) was considered to be such a candidate.

The sectional area and VCA data were modified to soften the stern shoulder, achieving the shape modification shown in Figure 7. No attempt was made to accurately match displacement or LCB, or to consider changes in wetted surface. The inflection was simply faired smoothly within the existing length of run. The influence of this change was solely to the ADVM prediction of wave-making drag. The viscous and form resistance components remained unchanged. (This is admittedly a simplification, one that was deemed conservative and justified as a cost-effective design optimization that would yield minor additional savings with a hydrostatically-correct re-analysis.)

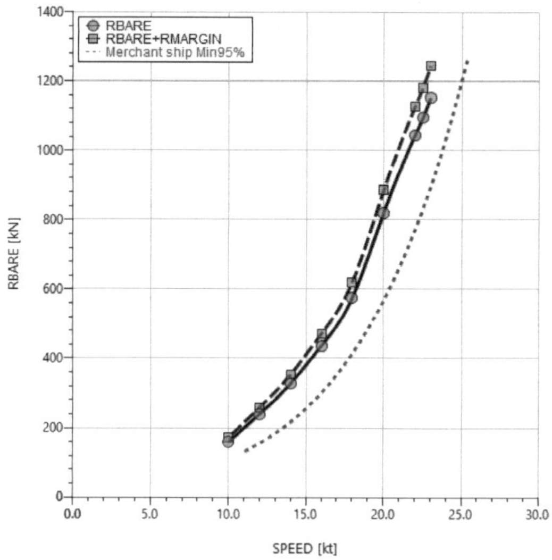

Figure 6: Comparison to "Confidence plot"

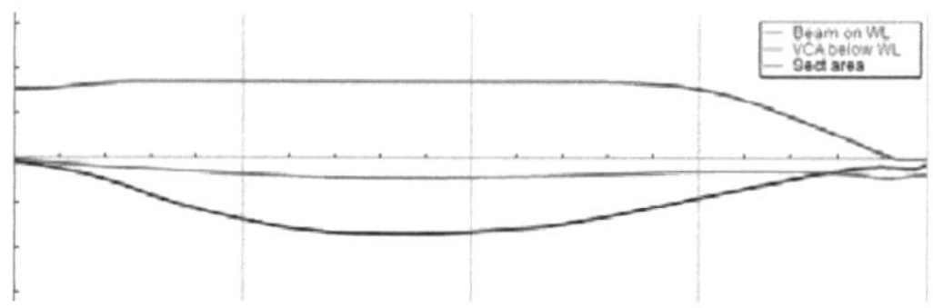

Figure 7: Illustration of stern modification

The effect of the stern modification on total bare-hull resistance was remarkable. The bare-hull resistance was reduced by more than 10% at the design speed (FN 0.31). The plot below in Figure 8 indicates the ADVM predictions of bare-hull drag for the original and smoothed stern shapes (in resistance/displacement ratio format for the constant displacement).

Anticipated follow-on work is to conduct a matrix study of stern options that include the effect of volume changes on the hydrostatic loading condition. For the purposes of this project, however, the resistance prediction for the original and modified hull forms – plus added resistance for appendages, wind and waves in sea-state 3, and a design margin – served as the basis for the thrust requirement for propulsion sizing and analysis. No additional iterations or higher-order computational or empirical testing was deemed justified at this stage in the design.

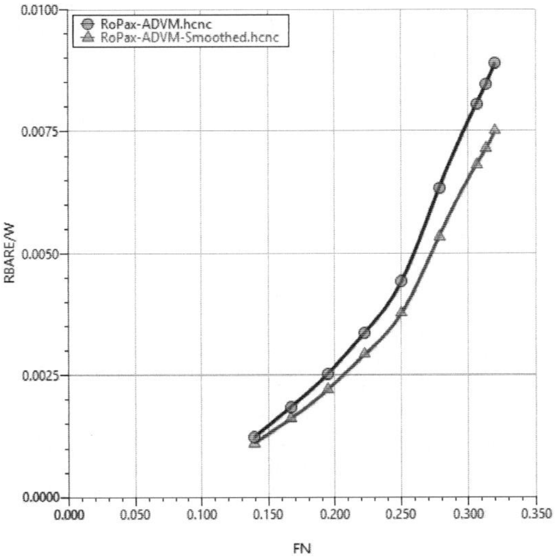

Figure 8: Bare-hull drag reduction with modifed stern

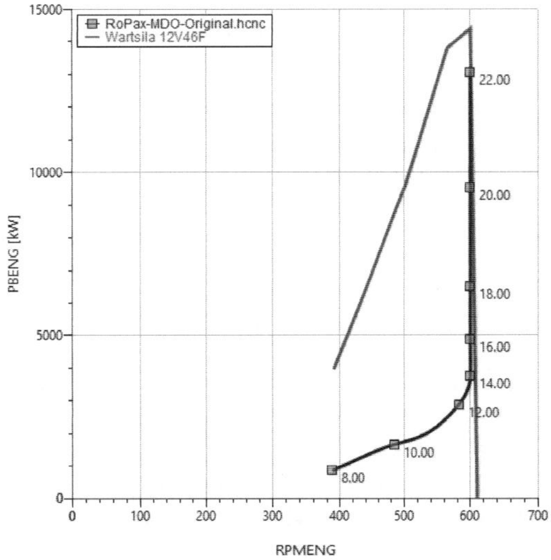

Figure 9: Engine (12V46F) loading with original hull

5. STEADY-STATE PROPULSION ANALYSIS

This twin-screw vessel was to be powered by marine diesel main engines, driving a typical single-stage non-reversing gearbox and controllable-pitch propellers (CPPs). The mode of CPP operation was specified to be constant RPM, with pitch varying to meet equilibrium powering load requirements. (RPM correction was allowed at low-speed modes to insure that CPP pitch remained above a minimum pitch setting.)

Likewise, the NavCad software was applied for the computational predictions for all propulsion analyses. This included prediction of hull-propulsor interaction coefficients (wake fraction, thrust deduction), equilibrium CPP pitch (for steady-state thrust matching at each speed and engine RPM), and all derivative propulsion metrics such as power, efficiencies, engine loading, fuel rate, CO2 production, and cavitation assessment.

5.1 COMPARATIVE EVALUATION OF INITIAL AND PROPOSED DESIGNS

The engine selected as the basis for comparison of fuel alternatives was the Wartsila model 12V46F (Main Engine specification), rated 14400 kW at 600 RPM. The engine power curve and fuel consumption was entered per the published product model literature. The following plot shows the model's published power curve overlaid with the propeller demand curve using the prediction with the original hull. (You will note the constant 600 engine RPM with low-speed correction for pitch constraint.) The power used is nominally 13100 kW per engine (or 91% MCR of the 14400 kW rating).

The proposed alternate engine was the same model in its dual-fuel configuration; a pair of Wartsila 12V46DF engines rated 13740 kW at 600 RPM. The engine load plot for this engine – with the modified stern – is shown below. The engine brake power used to achieve 22 knots cruise speed is approximately 11200 kW per engine (or 82% MCR of its 13740 kW rating).

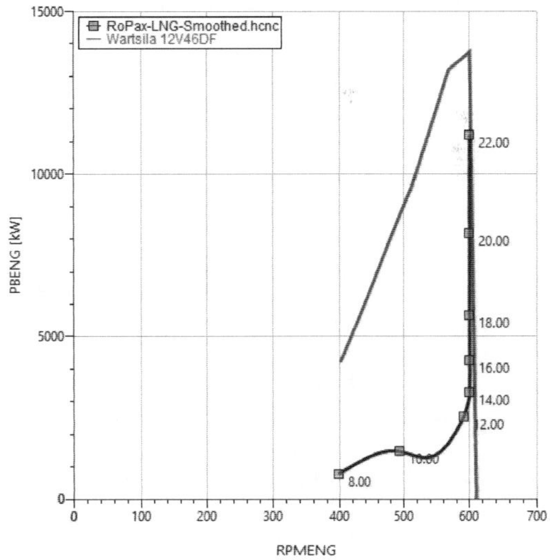

Figure 10: Proposed engine (12V46DF) loading with modified stern

It is worthwhile to also point out that there is a small additional reduction in power demand that come with the reduced loading on the propeller. The propeller design point for each condition was based on thrust matching at the design speed of 22 knots (NavCad's sizing "By total drag"). Propellers can be inherently more efficient with lighter thrust loading when the sizing is integrated into the propulsion analysis, and the objective functions allow for optimum sizing of blade area ratio and gearbox reduction ratio. As the final values for both were not yet defined, there is opportunity in early design stages to utilize a reduced BAR and an optimum commercially-available reduction ratio. The benefits of the lighter loading are modest (open propeller efficiency increases from 0.660 to 0.676 – equivalent to a 2.4% power reduction), but still of benefit to the reduction of GHG and fuel consumption.

6. OPERATING MODE STUDIES

Operating modes analyses were conducted for the two conditions. The results below show totals for the 354.6 nm one-way trip defined above in the duty-profile.

Condition	Original	Proposed
Fuel	MDO	MDO/LNG
Total fuel energy	3.10 MJ	2.66 MJ (-14%)
Total fuel mass	72.5 t	55.5 t (-23%)
Total fuel volume	80.5 m3	122.2 m3 (+52%)
Total CO2	31.7 t	20.9 t (-34%)

There are 288 such trips per year (144 round-trip voyages). Projected annual CO2 production would be reduced by approximately 3100 t.

7. CONCLUSIONS

Principal conclusions from the study are:

- The NavCad ADVM bare-hull resistance prediction method – as representative of intermediate [2D] methodologies – clearly demonstrates its success as a tool to help identify improvements in hull form, especially at early design stages and without the expense of higher-order methods.
- Significant bare-hull resistance reduction (of more than 10%) was achieved with smoothing of the sectional area curve at the stern run (from shoulder to transom).
- Small additional improvements in resistance might be achieved with a follow-on hull form optimization refinement using an iterative matrix of stern options with minor hydrostatic correction for the operational waterline.
- Reduction in thrust loading can provided opportunity for improvements in propeller efficiency if propeller BAR and gearbox reduction ratio can be selected as part of the sizing process. It is too late to take advantage of such benefits if the parameters are already selected and "locked in".
- Dual-fuel LNG engines offer substantial opportunities for reductions in CO2 GHG production, due to better heating (calorific) and carbon content properties than traditional liquid fuels (such as MDO).
- The compounded effects of reduced resistance (with a smoothed stern), better propeller efficiency, and lower power-specific CO2 production achieves the owner's original mandate to "reduce the vessel's carbon footprint" by the amount of some 3100 t per year.

8. ACKNOWLEDGEMENTS

The Authors would like to express their sincere "thank you" to all those who contributed with their work and efforts to this project, particularly to: Mr. Costis Stamboulelis, Executive Management Advisor, New Buildings Division of Blue Star Ferries, and Mr. Pratikakis, Managing Director NAP Engineering, for their continuous support during the project.

9. REFERENCES

References should be numbered in the text where they occur, with the number in square brackets [1] and a complete list given at the end of the paper.

1. IMO MEPC 281 (70). RESOLUTION MEPC.281(70) (Adopted on 28 October 2016) Amendments To The 2014 Guidelines On The Method Of Calculation Of The Attained Energy Efficiency Design Index (EEDI) For New Ships.

10. AUTHORS BIOGRAPHY

Donald MacPherson holds the current position of Technical Director of HydroComp, Inc., USA. He is responsible for all software development management and is principal investigator for hydrodynamic consulting services. A graduate of the Webb Institute of Naval Architecture, his previous experience includes positions in design offices and shipyards. Mr. MacPherson is a Fellow of the Society of Naval Architects and Marine Engineers, and member of its Propeller Hydrodynamics Panel.

Ioannis Bakas displays 20 years of solid Project Management experience, accumulated in blue chip companies like Hochtief AG, The Coca Cola Company and Lamda Development. He holds the position of Special Projects Manager at NAP Engineering, dealing with complete engineering projects on LNG as marine fuel on vessels, including retrofits and new building designs of ro-pax vessels, as well as design of LNG Bunker & Feeder Vessels. Furthermore, on behalf of NAP, he is responsible for running as Project Manager the POSEIDON MED II Small Scale LNG program which is co-funded by European Union. He is Dipl. Mechanical & Aeronautical Engineer MSc. NTUA, holding an MBA from Warwick Business School.

AUTHORS' CONTACT DETAILS

Jan-Erik Räsänen
Foreship Ltd
jan-erik.rasanen@forship.com

Gavin Allwright
International Windship Association
secretary@wind-ship.org

Jukka Kuuskoski
Norsepower Oy Ltd
Jukka.kuuskoski@norsepower.com

David Ferrer Desclaux
Bound 4 Blue S.L
dfd@bound4blue.com

Ken Goh
Knud E Hansen
keg@knudehansen.com

Steven Fletcher
Frazer-Nash Consultancy Ltd
s.fletcher@fnc.co.uk

Brent Perry
PBES
bperry@pbes.com

Petros Menegakis
Newcastle University
p.menegakis1@newcastle.ac.uk

Rogier Eggers
Maritime Research Institute Netherlands (MARIN)
r.eggers@marin.nl

Marios Ioannou
ABS
MIoannou@eagle.org

Martin Einsiedler
SHIPTEC AG
m.einsiedler@shiptec.ch

Donald MacPherson
HydroComp, Inc
Donald.macpherson@hydrocompinc.com

The Royal Institution of Naval Architects
8-9 Northumberland Street
London, WC2N 5DA
United Kingdom

ISBN 978-1-5108-8304-8

CURRAN ASSOCIATES INC.
proceedings.com